Design project planning

Frontispiece   An exquisite example of small mechanism design by SUWA SEIKOSHA CO. LTD. of Tokyo, Japan associated with a first attempt at design by my grandson Christopher, aged five.

Design project planning

A practical guide for beginners

W. T. F. BOND

PRENTICE HALL

London New York Toronto Sydney Tokyo Singapore
Madrid Mexico City Munich

First published 1996 by
Prentice Hall International (UK) Limited
Campus 400, Maylands Avenue
Hemel Hempstead
Hertfordshire, HP2 7EZ
A division of
Simon & Schuster International Group

Typeset in 10/12 pt Times by
Mathematical Composition Setters Ltd, Salisbury, Wiltshire, SP3 4UF.

Printed and bound in Great Britain by
T J Press (Padstow) Ltd

Library of Congress Cataloging-in-Publication Data

Bond, W. T. F.
   Design project planning / W. T. F. Bond.
     p.   cm.
   Includes bibliographical references and index.
   ISBN 0-13-349275-3 (alk. paper)
1. Design, Industrial. I. Title
TS171.B64 1995
745.2—dc20
                                95-14235
                                   CIP

British Library Cataloguing in Publication Data

A catalogue record for this book is available from
the British Library

ISBN 0-13-349275-3

1 2 3 4 5 00 99 98 97 96

I dedicate this book to my wife Rose
with my love and gratitude
for all she has done in helping
to make it all possible

W to R

Contents

List of plates

List of illustrations

List of examples and case studies

List of data sheets

Preface

Knowledge, experience and data are converted into products by thought processes unique to every designer – which may be described as his or her own style. As with all art forms style evolves as people learn to apply certain basic skills to those aspects of human behaviour that design education uses as a foundation – natural ability, enthusiasm, temperament, specialisation, academic studies etc;

There is an ancient Chinese proverb that translates roughly as follows:

> Tell me and I am interested
> Show me and I will remember
> Let me do it – and I will understand

Every experienced designer will support this philosophy and there is no doubt that the skills which contribute to what I have called 'style' are best acquired by personal involvement in design projects and real case studies. Thus, I have chosen to present much of the advice given in this book in the form of case studies, using these to introduce important and useful design techniques of primary interest to students attending courses which include the subjects engineering design or engineering product design, or those embarking on a design training programme in industry (in this book all interested readers will be referred to collectively as students). It is assumed that students at this level will have already gained experience in small free-thinking type projects and are now beginning to tackle complete projects – a brief followed by a design study and culminating in the manufacture of a prototype model.

Everything that is used by society, including all professional activities, depends on machines, equipment and apparatus which have been created by designers. These extend from machinery which processes raw materials and manufactured products to those which assemble, test and finish products. The spectrum of design activity is vast – but there are basic concepts applicable to all design projects and it is these that form the main focus of the book.

Whatever their background and training designers gradually evolve techniques and methods of working which suit each individual's instincts, skills and temperament. However one has to start somewhere! In Chapter 1 will be found advice about getting involved in practical aspects of the design process. Included here are techniques, thought processes and 'golden rules' contributing to a frame of mind developing an approach to designing.

Chapter 2 presents a case study to demonstrate an approach to designing which helps with handling the masses of information generated by the project, interpreting the brief or specification and establishing design objectives for the major features of the product.

The major features of many projects, particularly those concerned with the design of machinery, are the mechanisms around which the machine functions. The geometry may be devised with the aid of a drawing board or computer but, before expensive prototypes are constructed, it is frequently necessary to test the anticipated performance of a machine using experimental models (a process which is also a valuable source of creative ideas).

It is therefore of paramount importance that designers develop an understanding of the data that models will provide, together with the skills and visualisation necessary for making decisions on model design to ensure that they will provide adequate and usable data. An introduction to these objectives is the subject of Chapter 3. Here the reader will discover the reasons why the development of practical skills is of vital importance. Comments describing design and data feedback from models, advice on construction and much else that needs to be included in a designer's armoury of skills is also included. This chapter is written around a number of case studies and describes aspects of designing which are difficult to find in design literature. It will also partially endorse the aforesaid statement – 'Let me do it and I will understand!'

All these chapters are practical guidelines which will help those new to design studies avoid many of the pitfalls that lead to disappointing results, but which do not inhibit free thinking in project visualisation.

Another major objective of the book is the provision of project-planning techniques which can be used immediately by new designers but which does not demand an understanding of advanced techniques (the subtleties of which can only be properly appreciated with advanced experience).

During early design studies the time available for project work is often limited; another objective of this book is to demonstrate how good project planning enhances the flow of creative ideas and, in so doing, accelerates the design study process. Associated with advice given in the text are a selection of brief project specifications and guidelines in the form of data sheets. Example project specifications may whet a few appetites in the choice of project subjects or serve as catalysts for devising new project ideas; the guidelines are written to assist with the build-up of design knowledge and will be useful as a constant reference during early design activities. To facilitate easy reference the data sheets are placed in the last section of the book.

Readers should have no problems with the literary style used for this book although I have not hesitated in using terms which can be described as the language of the engineer. These terms become familiar in time and a glossary is available for those new to engineering design studies.

The book is about designing, and although written around my own main interest, special purpose machines, much of the advice given here will make a contribution to other types of training programme, where designers may apply the recommended techniques to their own field of design interest. For example, the engineering design case studies could be changed to those appropriate to another design discipline; the planning methods,'golden rules' applicable to industry and the general philosophy would remain virtually unchanged.

W.T.F. BOND

Acknowledgements

I would like to offer my sincere thanks to the following:

Rosalyn and Glenna for typing the manuscript and making such a good job of inter-preting my handwriting.

Fred Ayres for contributing DATA SHEET 15.

G. Wittenberg and V.J. Shend'ge for their interest and for contributing the folding machine model discussed in Chapter 3.

The Ministry of Defence for permission to use the photographs shown in Plates 1A and 1B: Prototype machine for air sampling and chemical analysis.

The Royal Aircraft Establishment for permission to use the photographs shown in Plates 2A and 2B: Prototype recorder.

Dr A. N. Barnes, Felpham, Bognor Regis, for the Frontispiece and Plate 3 photographs.

1

Project studies – getting involved

1.1 BASIC PRINCIPLES

Design courses and training programmes in design often begin with a number of free-thinking projects designed to exercise the imagination or to give practice in particular design skills. Project studies, in the context of this book, begin at a later stage when a project commences with an idea or a brief description of a product and where a design study is required for the purpose of establishing a product specification (later some of the research aspects of a design study involving prototype models will also be considered).

The degree of complexity of major design projects usually increases as the new designer's store of knowledge (practical, technical and scientific) is being extended. At an advanced stage a reasonable understanding is expected of what is meant by 'good design practice', but principles that form the basis for this also apply to projects undertaken early in a course of training and need to be implemented by new designers from the very beginning. Once these are thoroughly understood (one of the main objectives of this book) they will form a solid foundation for the quality of design project management demanded by industry.

The essence of good project management begins with the disciplines listed below; however, it must be appreciated by readers new to design that a full understanding of the *implications* of such rules can only come with experience over a period of time; but living with them, from the very beginning of a career in design, has important advantages, many of which will be explained later.

■ There is no substitute for a thorough practical training – seize every opportunity to become involved.
■ Good design results from efficient project planning – seek all available information about a project from the very beginning – study the specification in every minute detail (explained in Chapter 2).
■ Every component of a product is important – do not treat details casually.
■ Clear and concise communication with everybody associated with a project is vital.

- Drawings must contain all information necessary for manufacture and quality control.
- Confident application of the basic laws of science and mathematics comes only with adequate practice.
- A broad knowledge of manufacturing techniques is essential.
- Solutions to certain design problems can only be found by research using prototype models (explained in Chapter 3). An ability to specify the type and quality of prototypes is of the utmost importance in many industries.

These are matters which should be firmly established in the minds of new designers and will always be criteria for good design practice. Other 'golden rules' are concerned with customer relationships, many of which are associated with a particular type of design activity. (The Appendix, page 92, provides insight to the enormous range of activities undertaken by various types of designer.) There are three rules however, common to all design projects in industry and these must also be accepted as basic principles:

1. A product must be designed within the criteria detailed in a specification; any changes to these instructions must be negotiated and agreed by a customer before they are implemented (except where a contract allows discretion in the interpretation of design objectives).
2. The quoted costs of a design study should not be exceeded without prior agreement with a customer (excluding what are known as cost plus type contracts where some other form of financial control is agreed).
3. A design study must be completed on a contracted completion date (except where a contract allows flexibility due, for example, to uncertainties in time estimates associated with the need for a large design research effort).

These disciplines are often difficult to maintain in a training environment; designers in industry have the advantage of being surrounded by a commercial infrastructure focused on particular types of product (facilities, materials, staff support etc). However, new designers working in a college environment should always aspire to maintaining these three principles (for 'customer' read 'supervisor').

One other golden rule for new designers which can be considered basic to good design practice:

4. Design *everything* – all paperwork, reports, drawings, homework and coursework: nothing should ever be presented that has not been designed to the best of one's ability.

Many interactions between the principles listed here are explained in more detail throughout the book, and awareness of these will be extended by a designer's own experiences. They are all related, in various ways, to the production of creative ideas, the sources of which cannot be positively identified because so many thoughts occur virtually simultaneously when exploring a design problem. It is convenient therefore to consider separate elements of a designer's complex thought processes and we begin with the process of problem solving.

1.2   PROBLEM SOLVING

When solving problems part of a designer's brain seems to function like a three-dimensional spiders web; where each filament is a source of knowledge and where ideas generate 'vibrations' that are felt and influenced by the whole structure; a kind of resonance which continues until a viable solution is recognised. This is just one scenario into the working of the mind (and many others have been suggested) – a process which nobody can explain. Whatever the reality we can be sure that no problem is solved in isolation when the whole of life's experiences are brought to bear on finding a solution. However, in practice we do have to isolate *aspects* of problem solving in order that we may cultivate attitudes and approaches to designing and so that we can pass on experience.

We begin our journey into the mind of the designer by making an assertion that training and experience alone does not produce the best designers; other qualities are essential and need to be cultivated from the very beginning. Included amongst these qualities are:

1. A determined and tenacious approach to problem solving.
2. The type of mind that examines every aspect of a problem on a very broad and detailed basis.

These attributes, coupled with an enthusiastic interest, make a major contribution to the development of an inventive mind. What follows is a case study for a simple mechanical problem that is intended as an early 'glance' into the mind of a designer and a demonstration of the above statements put into practice.

EXAMPLE 1.1   Trapdoor mechanism

Figure 1.1 is a schematic side view of a conveyor belt system. Packages are transported by moving belts to processing areas 1 and 2 in the direction shown. Deflection of

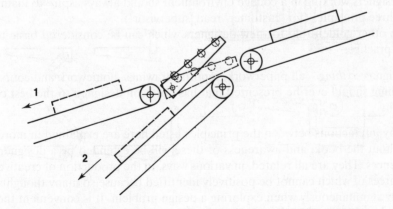

Figure 1.1   Conveyor schematic

packages from area 1 to area 2 is controlled by the position of a trapdoor. This is operated pneumatically by a device which counts the number of packages passing along the belt. This case study concerns the mechanism which operates the trapdoor.

It doesn't take the ingenuity of a designer to visualise an obvious solution to this problem – all one needs to do is put a lever at the trapdoor pivot and provide an air cylinder to push and pull the lever! In practice, this type of spontaneous reaction to a first idea leads to a process of repeated design modifications as further study reveals factors not yet considered; design failures often result from what may be described as a shallow treatment of the factors *influencing* the solution of a design problem. This doesn't mean that immediate inspiration, born of intuition, hasn't a role to play in problem solving; a flash of insight can produce a first-class solution to a problem – but the experienced designer will make a few notes and sketches of such ideas and then pause to consider other features of the project which may influence his solution. For the trapdoor mechanism typical data is listed in Table 1.1 and a few outline concepts of possible mechanical assemblies are shown schematically in Figure 1.2 (there are others but those drawn are sufficient to establish the point of this case study).

It is obvious that many of these features would be familiar to an experienced designer and these would be dealt with almost immediately; one or two will identify decisions to be made and thus be pondered over and discussed while others may need further communication with associates or equipment suppliers. The point to be made is that

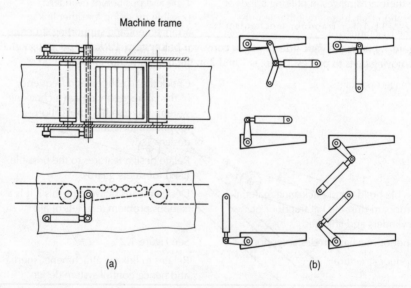

Machine frame

(a)   (b)

Figure 1.2   Trapdoor schematic. (a) Schematic plan and sideview of trapdoor showing position of air cylinders and links. (b) Alterative layout of components.

Table 1.1   Considerations for the trapdoor design

Feature	Design implications
Velocity of packages	Speed of trapdoor operation is related to the type of fixing that could be used at the trapdoor pivots (pin, keyway, spline etc) also pivot dimensions and linkage design
Weight and size of packages	Geometry of structure beneath the trapdoor to avoid damage to deflected packages
Weight of trapdoor	Type and size of air cylinders: design features of pivots and linkages
Power source details	Convenient access to air supply: best layout for pipe run: positioning of air cylinders
Trapdoor actuators or air cylinder types and mechanical linkage design	Position and design of machine structural features for support, access and maintenance
Reliability of conveyor belt velocity and spacing between packages	Package arriving at trapdoor out of phase – possible need to investigate fail-safe features for trapdoor
Are there advantages in utilising trapdoor operation to actuate any other feature of the machine?	Type and position of cylinders: possible provision for other linkage arrangement and supporting structure
Possible overload of packages at area 1 or 2	Possible need for feedback signal to pneumatic control circuit
Damaged packages	Check the likelihood and frequency of this: if necessary design a checking system: this also has implications for pneumatic control circuit design
Safety	Relate design features to the possible need for safety guards
Possible build up of packaging material (cardboard dust etc.) at trapdoor or at cylinders and linkages	Design means of avoiding this if it is a serious problem (e.g. jets of air)
Trapdoor mechanism design options	See Figure 1.2
Trapdoor structure	Relates to linkage attachments, inertia and hence control system design

all available data is considered *before* making final decisions on the mechanism design. This simple example demonstrates that examining every aspect of a problem involves utilising the whole of one's experiences, feeding various combinations of ideas into a set of acceptable parameters until a satisfactory solution emerges. Thus, a whole armoury of possibilities is mentally scanned – not just the first couple of ideas that occur. Developing an ability to scan an array of possible solutions is particularly useful when a designer is faced with what appears to be an intractable problem. It is then that a determined and tenacious approach can often prove fruitful; one must be absolutely determined that a solution will be found.

This case study has included some thoughts about a number of mechanical arrangements; as time passes the new designer will rapidly increase the store of knowledge relating to mechanisms and these, with all their combinations and variations, provide an immense reservoir of potential solutions to mechanical problems; the primary objective of the above case study, however, is to explain some of the *thought processes* associated with problem solving, to demonstrate that a broad examination of a design problem results in a useful choice of options. We now examine how this philosophy influences thoughts about a simple project brief; but first a few other pertinent facts.

A thought which enters the mind of most students after writing, or being presented with, a project brief is 'where do I start?' For those who are unfamiliar with design, the notion that a project begins as an information-gathering exercise comes a surprise.

Once a project subject had been chosen there is usually an enthusiastic drive to 'get on with it' and to begin drawing various solutions. Enthusiasm is good and is to be encouraged – but minimising the investigation phase often leads to bizarre, impractical or over-ambitious solutions (time and money are always limited and it is obviously necessary to concentrate on concepts which are most likely to produce a viable solution).

Designing is largely a process of decision making and a useful method of using the investigation process to generate the right decisions is by writing a list of all general factors which are thought to be relevant to the project. Such a list will inevitably reveal a number of queries which can be tackled immediately – a process which generates decisions, reduces the number of design options and begins to eliminate unlikely solutions; such a list will also reveal that virtually all factors can be categorised in two ways:

1. matters concerned with design features – for example early ideas on method of operation, mechanisms, weight, size, aesthetic and ergonomic considerations etc.;
2. matters relating to what can be defined as design organisation – amount of time and money available, access to facilities, essential visits to outside establishments, completion date, identifying essential research programmes, etc.

Enthusiasm tends to give priority to category 1 features, but there are many advantages in early consideration of category 2 matters; it is these aspects of project planning which are often neglected until some minor crisis forces attention toward them. Professional designers always advise students to make a very thorough and careful study of design organisation factors at the very beginning of a project, to spend adequate time on this task and to keep neat records for future reference.

Influences of both categories are discussed in Chapter 2, where a case study demonstrates project planning for a relatively complex project. With less complex projects, however, a simpler procedure can be adequate for the purpose of getting involved. Project planning is an information-handling process, the usefulness of which can only be appreciated by observing the concepts that it generates; to this end we begin an introduction to the techniques involved by referring to the following brief:

EXAMPLE 1.2   Grip trainer

The care and management of disablement in children often presents a need for special-purpose apparatus to assist with various types of physiotherapy. One such apparatus may be described as a grip trainer, a device which provides a visual, audible or edible 'reward' when a preset gripping force is achieved by the patient.

This type of therapy often follows injury to the hand and muscle strength is improved by gradually increasing the gripping force required to achieve the reward. This force varies with each patient and therefore must be set and adjusted (very gradually) by the physiotherapist, which implies some form of force indicator.

A prototype is intended for the use by children in the four to six age range, which specifies a gripping force varying from very light – virtually no reaction force – to that which can be achieved by a normal 6-year-old child.

Assignment

(a) Design a prototype incorporating one or more of the types of reward suggested.
(b) Complete your design using neat freehand sketches.
(c) Manufacture a prototype.
(d) Prepare a very brief critical appraisal of your product relating to:
  (i) areas of particular difficulty in manufacture;
  (ii) design features which would need modifying before a mark 2 prototype could be manufactured.

Several tasks are included in the assignment for the benefit of those who would like to attempt this project; we will concentrate on the project planning exercise for (a).

I prefer to begin a project-planning study with a bit of free thinking about the product. It is to be admitted that after raising a few queries with a customer subsequently, some aspects of a free-thinking phase may prove to have been unproductive – since one comment alone from the customer can change a preconceived idea. But, with an open-ended specification such as this, a short time devoted to your own concepts of possible solutions, before the ideas of others prejudice the mind, often produces useful new ideas. Also, it is my own experience that if later queries rule out your early concepts of an overall solution, time has not been entirely wasted. This is because the thought processes of an experienced designer sweep over a whole spectrum of relevant parameters – which tends to prepare the mind to be even more receptive and diagnostic at a later stage, as information emerges from other sources.

Thus we begin by carefully analysing the specification and listing major design features together with comments that might be useful. This is where imagination about the project details really begins, by visualising each feature actually located or functioning. For example, let us look at the action of gripping by visualising the following:

- a child squeezing a device;
- the grip opening and closing;
- the device located close to a reward system, attached to a reward system, or remote from it;
- hands of different sizes holding the device;
- the relationship between the tips of the fingers and the palm of the hand, with the squeezing action;
- attaching the device to the reward system (wire, rod, cable) and its location relative to the fingers;
- the device fixed at various points – centre, side or end;
- weak points: where breakage could occur through fatigue;
- the gripping device actually fixed to a reward system (thus eliminating the need for a connection method);
- the shape of the device held comfortably in the hand (made from a variety of materials);
- etc.

The following concept is being conveyed: when thinking about a feature *exhaust all design possibilities emerging from your experiences*. These concepts take a lot of time to explain. In practice the designer's mind works at high speed, with ideas emerging (and being accepted or rejected) in seconds. But one must make the effort to do it.

Incidentally, you will find that ideas are being produced in your mind even when not working on the project – particularly when you have experienced the procedures recommended here.

Referring again to our project, the first step may look something like that given in the lists below; such lists will never contain everything and designing is an evolutionary process, with later ideas suggesting even more ideas (a process which gradually diminishes as a project progresses and major design features are established). Hence, from a free-thinking exercise based on the specification, we arrive at the following ideas:

Item – visual reward

- use flashing or moving lights;
- grip action switches on a colourful computer display;
- activate an articulated figure (plywood?);
- activate a toy robot;
- adapt something that is already made (torch, toy car, etc.);
- design a moving coloured mechanism;
- use a 'surprise' type of reward (unexpected event);

- use combinations of any of these suggestions;
- reward could be operated by mechanical switch, contact switch or microswitch;
- time period for reward? must it switch off when grip is released or with a delayed action? – discuss this with therapist;
- if lights are used ensure easy replacement of bulbs.

Item – audible reward

- utilise a music box;
- a music box would also provide a visual reward;
- look around the toy shops;
- switch on radio, television or favourite tape – discuss using familiar audible rewards with therapist (are these ideas viable?);
- talk to other young children for their opinions.

Item – edible reward

- are any particular sweets preferred? (talk to therapist);
- sweets could be fixed to something which rotates;
- sweets of different sizes could be held in a small box (or boxes) with a trapdoor release;
- if a 'vending machine' type dispenser is used sweets of the same size would be easier to handle;
- a reward sweet could follow a path around objects in a vertical 'window' (also a visual reward) it could even switch on lights – possibly using a simple switching circuit or light sensor;
- sweet could pop up or out of something;
- edible reward could be linked to visual reward by a toy presenting a sweet to the child (puppet?).

Item – gripping

- allow for both hands;
- measure width of childrens hands;
- device may need adjusting for hand size;
- check start of grip – is it with hands half open?
- grip by squeezing only – grip basic shape must not change with strength adjustment (talk with therapist);
- note everything mentioned earlier in the text (concentrated thoughts about the gripping action).

Item – setting the gripping force

- rotate a large knob (buy one?);
- push a lever;

- provide a scale or dial (an existing component could be used if it is calibrated to the grip force);
- investigate the possibility of using one spring to set the whole force range;
- design a method for measuring the gripping force of a 6-year-old child (is a suitable pressure gauge available in the science labs?);
- investigate other ways of obtaining the gripping force (bending a lever or rod?).

Item – other ideas

- gripping unit – fixed or movable ? – talk to therapist;
- talk to therapist about overall image – two ideas: gun with gripping action to 'fire' the reward – toy rocket launcher;
- it might be encouraging to a child to observe a pointer as the grip is increased – reaching a mark to achieve the reward?
- investigate using a force transducer in the gripping device;
- where will apparatus be stored – will a protective case be required?
- battery or low voltage power supply (what electric supply is allowed?);
- perhaps a wind-up spring could be used to activate the reward;
- it may not be difficult to offer all three types of reward (therapist would have access to hidden selection switches) – talk to therapist and relate to project time estimates later.

Every student undertaking this project would write a different list of factors emerging from the brief specification (although each list would have factors in common with others). The content of such a list depends on the experience of the student, any previous knowledge of the project, the time, money, materials and workshop facilities available. Other influences will also be present in the mind when making decisions, such as the following;

1. Personal interest – the desire to associate the project with a personal hobby or a preconceived outline solution.
2. Professional guidance – a strong request from the Physiotherapist for a particular solution to the reward system.
3. Advice of supervisor – who, by experience, can estimate when a suggested solution is under- or over-ambitious.

Such influences as these are usually associated with a project at the beginning of the design study and therefore are automatically integrated with all other considerations, thus focusing thoughts on a *narrower range of options* than those listed above.

When questions have been answered and the shopping around completed, your original free-thinking list will have been shortened considerably and you will now be sketching possible solutions. It is here that you need to pause and write another brief list of the *major subject areas* associated with this project and question how adequately these have been considered:

1. *Medical factors*

(a) Is there anything other than the gripping and reward features which need to be considered e.g. how long before the patient gets bored with the reward – it might be possible to incorporate several (selectable) actions from the same set-up?

(b) Will the apparatus be used by disabled children?

(c) There must be a temptation for the patient to use the other hand to obtain a reward – this could be prevented by providing a 'master' switching device to be activated by the other hand.

(d) Is there a need for an instruction plate fixed to apparatus (other users or departments)?

2. *Ergonomics*

(a) The gripping action must be comfortable – rounded corners and edges.

(b) Relationship between height of patient and table height? or could a device stand on the floor?

(c) Design gripping device so that pinching of the hand or fingers is impossible.

(d) If the apparatus is to be located on a table ensure that the patient cannot pull it off with misuse of the gripping device.

3. *Aesthetics*

Seek an aesthetic response in the child – a motivation feature? talk to therapist.

4. *Reliability*

Able to withstand periodic knocks; reliable mechanisms (avoiding need for lubrication); provide for periodic testing of grip force.

5. *Materials*

Avoid materials which corrode easily, thin sections (design for long life), scuffing of moving parts.

6. *Manufacturing*

Design for easy maintenance (children have a habit of posting small articles through holes and slots).

7. *Science*

Establish a calibration method for gripping force (if necessary).

8. *Safety*

Top priority; use non-toxic finishes.

9. *Models*

Consider what might need to be studied in model form before committing an apparatus design to precise manufacture.

When most queries have been dealt with and one begins to visualise and sketch possible solutions, it is useful to think of the end product as a closed system which has associated *input, activity, output,* and *waste* features.

Definitions relating to these features are:

- *input* – anything that is done to the product;
- *activity* – anything happening to the product while in use;
- *output* – anything emerging from the product;
- *waste* – anything detrimental to the product.

These features are interpreted broadly while visualising the product from every viewpoint.

The grip trainer is an unfamiliar product and further work on this will be left to those who may wish to use it as a formal project. It is more convenient to demonstrate the 'closed system' technique using a familiar product:

EXAMPLE 1.3   Washing machine

Imagine that a free-thinking phase has been completed (as with the grip trainer) and, accepting that some of the features listed below may have already been considered, a closed system approach often produces a few useful ideas and completes a very thorough investigation phase as follows:

- *Input*
  —access to the machine
  —method of connecting to power supply
  —leads and fittings
  —method of connecting to water supply
  —transporting the machine
  —storing the machine
  —dial and switch design
  —ergonomic features
  —aesthetic features
  —packaging
  —simplicity and quality of instruction pamphlets

- *Activity*
  —load capacity
  —heat dissipation to the environment
  —vibration (noise)
  —motor cooling?
  —mechanical and electrical safety features
  —detergent quantities required
  —washing efficiency
  —sealing (long-term efficiency)

- *Output*
  —removing the washing
  —cleaning the machine
  —access for maintenance
  —steam output to the environment
  —labels
  —finishing (the machine presents an image)

- *Waste*
  —power consumption
  —corrosion

—wear rates of catches, handles, hinges, castors
—wear rates of brakes and microswitches
—belt drive wearing
—dust build up
—anti-vibration fittings.

Obviously, many of these parameters would be considered at an earlier stage of the design study – but a *concentrated* effort, as demonstrated above, always pays dividends and takes very little time. It is also possible that parameters may have been missed in spite of these efforts; but each of the features listed above will be studied in detail later in the design study and any missed items usually become highlighted then (parameters relating to the internal mechanisms and control systems of the machine would be tackled concurrently with the above). There are no rules in implementing these techniques – they are just aids to the thinking process.

In conclusion the following three steps have been recommended

1. a free-thinking phase – visualising each main feature functioning and listing ideas and queries;
2. concentrating on main subject areas and questioning how adequately these have been considered;
3. when questions have been answered, and solutions begin to emerge, think of the product as a closed system in terms of input, output, activity and waste.

There is no doubt that these simple techniques result in ideas, new thoughts, things to do immediately, early communication with the customer, getting the feel of the project and, of immense importance, a beginning to the process of forming mental images (and probably a few sketches of possible solutions). In practice designers will rapidly develop methods of working that suit each individual's temperament and emerging skills; this is a good way to start!

In presenting these examples I have taken licence in postponing discussion on some major issues which play a dominating role in projects undertaken in a commercial situation. This has allowed discussion to be confined to thought processes associated with early projects undertaken by people new to design activities. In practice the choice of scheme for a project is influenced by many parameters not yet mentioned; including costs, completion date, required degree of quality etc. This requires a more detailed method of information handling which will be discussed in Chapter 2. Before this, however, guidelines for getting involved with other practical matters will be introduced.

Additional note. It is of interest that the thought processes explained in this chapter can have other useful applications. One example of this is given in Data Sheet 1 relating to the subject of packaging (which is not always outside a designer's responsibility – particularly in many thousands of small companies where the designer is often the focus of all activity concerning product design).

1.3   MANUFACTURING PROCESSES

The amount of time available for gaining experience in manufacturing processes is relatively short on most design courses; an introduction to elementary workshop or studio practices is usually within the experience of most readers qualified to join such courses, but a rapid increase in knowledge of manufacturing processes has a number of immediate advantages; for example, enhanced visualisation in design details (component shape, tolerances, assembly, finishes etc.) and an improved ability to specify components and to manufacture prototypes; all project work benefits in a subtle way if at least an early working knowledge of more advanced practical work is acquired quickly as a back-up to project studies.

One useful method of getting involved rapidly is a reading programme using a check list of commonly used processes such as those listed in Data Sheets 2 and 3. Many books on manufacturing processes are available and a working knowledge of these can be achieved quite quickly if the reader concentrates on the processes only; technical details can be assimilated at a later date or when those that are considered essential are introduced later in a course.

One other method of acquiring manufacturing knowledge quickly is by actually observing manufacturing processes; the value of industrial visits early in a design course is enormous and students are advised never to miss an opportunity to attend these.

Getting involved with manufacturing techniques early in a design career has important advantages. From the vast reservoir of possible examples, I can explain just a few – because many are related to particular industries and knowledge of these will come with later experiences. The following should be enough to convince that manufacturing knowledge can have considerable influence on the kind of design decisions which need to be taken at the drawing board or computer; they should also provide the motivation for efforts needed to acquire manufacturing know-how as a matter of some urgency. The *actual design* of components, in all types of industry involving some form of manufacture, is influenced by the *method* of manufacture. Confident decisions on component design (and indeed many general design parameters) can therefore only be achieved by accumulating a broad knowledge of manufacturing processes. Here are just a few examples to explain why such knowledge is vital:

1. There are preferred methods of manufacture for producing components in various quantities – single components, small batch, large batch or mass production. The quantity to be produced influences decisions such as tolerances, methods of holding and machining, jigs and tooling, quality control, choice of materials related to production methods and a a great deal more.

2. The design of a component (coupled with knowledge of quantities to be produced) dictates the sort of tooling required for manufacture – and hence tooling costs, which can be formidable when press tools, die casting or moulding tools are to be used. Some types of holding and positioning jigs are also very expensive to produce; these are important economic factors which are directly related to decisions taken by the designer.

3. Many industries are subcontractors to other companies. The components that they

supply are manufactured to a designer's drawings; therefore the designer must be aware of the methods of manufacture to be used, affecting subcontracting costs, component reliability, delivery schedules etc. (factors which influence the market viability of a product.

4. There are occasions when a choice of production methods is available to the designer; the skills and machining capacity within one's own company can influence design decisions (whether to use a casting or fabrication technique for example).

5. Some components need to be manufactured by a particular sequence of machining operations – all of which must be known to the designer of the components.

6. Industries work to manufacturing standards for tolerances, interchangeability, quality control methods, reliability and component finishes, etc.; these are just a few of the many processes associated with manufacture which must be reflected in decisions on component design.

7. The manufacturing method dictates the degree of accuracy achievable. In general, the higher the degree of accuracy required, the higher the cost (except for highly specialised manufacture where machines have been developed for very special work, e.g. electronic and optical components, precision gears, ball races etc.). One straightforward example is a small component which has to be mass produced by a casting technique. The designer needs to know the degree of accuracy achievable by the many available casting methods (see Data Sheet 3, Forming and treatment processes).

8. There is often a link between a manufacturing method and the ultimate method of quality control (inspection) to be used. This can involve decisions on component design related to allowances for special datum faces, or reference features for measurement checking (particularly where components need to be made with great accuracy).

9. There is a relationship between the method of manufacture and a subsequent method of assembly. If components are to be automatically assembled by special machines or robots, the designer must obviously be aware of the advantages of providing component features and allowances for any limitations or safeguards demanded by the manufacturing process.

10. A relationship exists between the method of manufacture and/or assembly and the type and quality of finishing processes (painting, plating etc.). Most of us are aware that without special processes (filling, polishing, buffing etc.) a good finish cannot be obtained by applying a finishing material to a poor surface. With component manufacture poor surfaces are identified with features such as tool cutting marks, grinding scratches, porosity of materials, scoring marks (through rolling processes) etc. It is therefore important that the designer understands and considers these manufacturing processes at the *component design* phase of his project.

11. There is also a relationship between the finished quality of manufacturing processes and the feel or appearance of a product – two areas of interest which demand the constant attention of the designer.

12. Relationships also exist which link manufacturing methods to many other design considerations, such as choice of materials, stress loading, potential for corrosion, wear

rates, durability, fatigue life and others too numerous to mention here.

Just a few reasons have been chosen to explain why manufacturing knowledge is an integral and vital aspect of project management ability.

Products are born from design concepts – not only the overall idea but concepts about every component and the way in which they assemble and blend together. The more knowledge one obtains about manufacturing techniques the richer becomes what I define as the 'concept resources' of the mind; it can never be too early to get involved with manufacturing processes.

Information listed on Data Sheets 2 and 3 can be obtained from books, films, videos, workshop demonstrations and visits to companies engaged in the manufacturing of product.

Students are advised to:

1. delete any process on the lists that are already sufficiently understood;
2. add to the lists any processes that you feel ought to have been included;
3. add to the list processes that emerge from new technology or special interests in a particular design field of activity.

Familiarisation with all of these processes will take some time but persistency in the acquisition of manufacturing knowledge will erode these lists surprisingly quickly.

1.4   TIME ESTIMATES

An organisation factor associated with all projects is time estimation. In industry this is a highly important activity and, in some companies, demands the work of a whole department, who use many complex procedures for information retrieval, calculation, manufacturing data, etc. It is obvious that the ability to fully appreciate such procedures can only come with experience, since reliable estimates depend on an ability to assess the time required to manufacture components and to assemble, finish and test a product. These skills will all develop with practice, but for new designers Example 1.4, given below:

1. provides a useful technique for scheduling early project work;
2. assists with the process of gaining the sort of experience that leads to improvement in estimating time factors;
3. serves as a support for greater understanding when more advanced techniques are introduced.

When the planning phase of a project is virtually finished and the brief converted into a more detailed specification a number of possible solutions begin to emerge in the form of rough sketches. At this stage it is possible to estimate the types of components and probable quantities required to manufacture a prototype even if these are not formally detailed, by using the following procedure:

EXAMPLE 1.4   Tricycle

Imagine that the project involves designing a small lightweight tricycle for three year old children. Steering will be controlled by a parent holding a steering rod attached to a swivelling back axle; the child is only required to pedal.

Assume that the designer is required to produce a prototype and wishes to estimate the amount of time needed to manufacture and assemble the components.

Each student will have received a different practical training and this will obviously influence time estimates. However, to begin with, a large overall contingency time factor can be used to allow for errors in estimating allowances for component manufacture through skill limitations, availability of machines and staff assistance etc.

The process involves making a list of components derived from early sketches and your estimated times as shown in Table 1.2.

To obtain an estimate for the whole project, time for drawings, evaluation and preparing a report need to be included.

Obviously, a similar technique can be used when experimental models rather than prototypes are planned.

If *actual* times for manufacture are carefully recorded these can be compared with the rough estimate; the valuable experience gained will result in an improvement in any subsequent estimates and also promote greater understanding for all estimating procedures.

This sort of activity is not divorced from estimates that need to be made in industry.

Table 1.2   Construction breakdown for Example 1.4

Component	Making time (hours)
Handlebar piece	2
Handlebar support bracket	2
Front wheel side supports (2)	4
Side support spacers (2)	4
Front wheel axle	2
Front wheel structure attachment	2
Complete seat unit	6
Two seat unit support brackets	4
Backrest	2
Backrest side supports (2)	4
Back axle	2
Steering rod to axle stirrup	4
Steering rod	2
Steering rod handle	2
Assembly and miscellaneous	6
Finishing	6
	54
Contingency (50%)	27
Total	81

True working rate – 10 hours per week
Therefore time estimate for manufacture = 8 weeks

Designers are constantly estimating time schedules and costs (of which more later) and constant practice from the beginning will provide the sort of experience that generates a high degree of skill in this important function later in life.

1.5   COSTING

Whatever the field of design it is essential that an appreciation of costs, relating to each design project, is integrated with other design activities; an ability to make cost estimates is vital to any commercial design activity and practice at this should be undertaken at every opportunity.

Costing procedures in industry are often highly complex, with whole departments engaged in this important activity. Obviously, a detailed study of costing methods is outside the scope of this book; but there are a number of simple techniques that can be employed by a student to begin getting involved.

As with time estimates it is never too early to practice cost estimation; even when a project is complete and a final check on costs proves to be considerably different from an early rough estimate, the exercise will provide valuable experience, some of which will be unattainable by any other means – such as developing an ability to constantly monitor (mentally) the likely effects on costs of changes in design options.

First it is necessary to accumulate a price list of commonly used items (screws, ball races, motors etc.); even when such items vary in price from time to time such records serve as a pricing resource for cost estimation. With a little determination and the cooperation of colleagues, such a record can be established very quickly. To this record can be added information provided by the college purchasing department (such as the bulk price paid for raw materials – fabrics, wood, metals, plastics, etc.) and manufacturers' price lists; it will also be necessary to (1) assume pay rates for the designer and, where appropriate, assistants (2) estimate the time required for various phases of the project and (3) assume overhead and profit margin figures. (Supervisors, consultants or designers working in industry are likely to be able to advise on typical percentages; these are never the same for any two companies but rough estimates are good enough for gaining practice and experience.)

Since there is an element of the unknown or the unpredictable about any new design venture, the designer must also learn when and how to allow for contingencies in quoting a cost. A contingency allowance depends on many factors, such as size and extent of the project, quantities to be produced, knowledge of the particular type of product, market forces etc. – all of which are assessed by experience; this is why cost estimation should always be practised in any design project situation.

For the student of design, costing also helps to identify which design concepts have sensible commercial viability.

Example 1.5 (given below) describes a method of making cost estimates which has the following advantages:

1. It is simple to use and to up-date.
2. It assists with the developments of estimating skills.
3. It serves as a 'primer' for advanced costing techniques.

EXAMPLE 1.5   Cost estimate

When a project has been planned to the point where a reasonably firm specification can be written and when a few sketches of a product emerge, complete the cost estimation form shown in Data Sheet 4. Use information pertinent to a commercial, rather than college design environment, even to the extent of estimating what assistance may be given by others. Assume this has been done for a project as shown in Table 1.3.

Table 1.3   Cost estimation

Item	Time (h)
Design study	10
Models and experiments	20
Formal drawings	12
Prototype manufacture	30
Modification	6
Formal report	10
Assistant effort	8
Assumed costing rates:	
Designer	£500/36 hr week
Assistant	£250/36 hr week
Overheads	200% (a typical figure)
Materials	£62
Profit margin	20%

Data sheet 4 is then completed as shown in Figure 1.3
Note: Calculations are rounded upwards to the nearest pound

Making rough cost estimates is not always straightforward, possibly because certain aspects of a design are difficult to visualise from a brief specification. We might, for instance, wish to quote an approximate cost for early discussions with a customer, rather than the estimated fixed cost of £4674. We can then make further estimates based on various assumptions by referring to the original rough estimates and the simple manipulation of a few figures on the calculator, as follows:

1. A contingency on labour costs of plus 6 h for the designer and 2 h for the assistant would result in an extra £229. One could then quote an approximate cost of (say) £5000 for the purpose of discussion with the customer.
2. A contingency of 100% in the estimate for material costs would result in a final cost of £4743 (an increase of £69 which might easily be absorbed in a contingency allowance).
3. An overall contingency of 10% on the total of £4674, would result in a cost of £5142.
4. Effects of various overheads can be examined:
   (a) 160% overheads instead of 200% results in a total cost of £4055.
   (b) 240% overheads instead of 200% results in a total cost £5288.

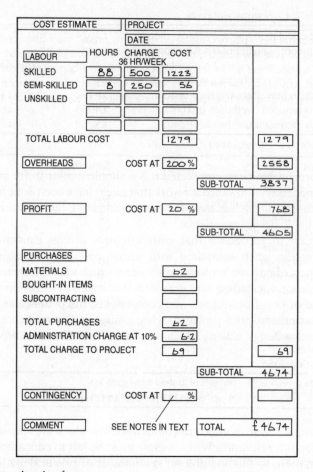

COST ESTIMATE	PROJECT			
	DATE			

LABOUR	HOURS	CHARGE 36 HR/WEEK	COST	
SKILLED	88	500	1223	
SEMI-SKILLED	8	250	56	
UNSKILLED				
TOTAL LABOUR COST			1279	1279

OVERHEADS		COST AT 200 %		2558
			SUB-TOTAL	3837

PROFIT		COST AT 20 %		768
			SUB-TOTAL	4605

PURCHASES			
MATERIALS		62	
BOUGHT-IN ITEMS			
SUBCONTRACTING			
TOTAL PURCHASES		62	
ADMINISTRATION CHARGE AT 10%		6·2	
TOTAL CHARGE TO PROJECT		69	69

		SUB-TOTAL	4674

CONTINGENCY	COST AT	%	

COMMENT	SEE NOTES IN TEXT	TOTAL	£ 4674

Figure 1.3   Cost estimation form

Let us summarise in Table 1.4 what has been calculated for this imaginary example:

Notice how an awareness for the cost effects of changing parameters emerge from such costing exercises.

It must be emphasised that costing for manufacture is a highly specialised procedure in the commercial situation; but making rough estimates by the simple technique described here is always possible and always provides valuable experience.

The estimate can also be calculated in reverse when a customer asks if the same project could be undertaken for (say) £3800. In this event we need to calculate restrictions on labour hours available to the designer if other parameters cannot be changed; this works out to about 70 h. We would then have to decide whether this is a viable proposition (possibly – if the customer did not insist on formal drawings or report). Thus are projects often negotiated.

Table 1.4   Summary of costs

First calculation (using 200% overheads)	£4674	
With contingencies of 6 h (Designer)		
2 h (Assistant)	£4893	i.e. +£219
With 100% contingency on materials	£4743	+£69
With an overall contingency of 10%	£5142	+£468
With reduced overheads of 160%	£4055	−£619
With increased overheads of 240%	£5288	+£614

When the project has been completed it is a simple matter to fill in another estimate (making necessary allowance for work that might have been done by others, such as detail drawings) using *actual recorded* time. Comparing this with the original estimation is a valuable design experience.

It is important to appreciate that early attempts at cost estimation are likely to be disappointing when compared with actual costs; but acquiring estimating skills could be described as an evolutionary process which improves rapidly with experience, and the process described here is a useful starting point since it involves learning about the price of bought-out items and components, discovering the way costs are influenced by overheads and profit margins, and gaining experience (without too much effort) into a design activity which is so vital to commercial success in product design.

1.6   COMMUNICATION

All types of designer have methods of communication which are special to their own particular design interests and advice on these must be left to educators in each discipline; however, there is common ground in a number of practical matters, which can be studied to advantage by all new designers concerned with presentation of work by lectures, displays and exhibitions.

In college the presentation of design work often begins during the first year of a course, continues throughout the course and frequently culminates in a degree show – where communication plays a vital role in conveying a student's level of attainment to examiners and potential employers. In industry these activities continue with the presentation of ideas and products to professional colleagues, customers and by exhibition at trade fairs, seminars and conferences etc.

Lecturing and exhibiting are activities which are repeated quite often on a course of instruction; for this reason and to facilitate easy reference, guidelines on these activities are given in Data Sheets 5 and 6. The latter advises on a relatively large exhibition but it contains elements which will also serve as a reminder for smaller displays.

Another form of communication that merits special mention is the design specification.

A product design begins with a communication which is called a *specification* – a set of instructions and/or intentions. The amount of information contained in a specification varies with the type of project and with large projects may require a document containing hundreds of pages. It may include agreements on function, manufacture, testing procedures, installation, future maintenance, servicing arrangements and much else.

The first lesson we must learn from this form of communication is that after a contract is signed the designer is obligated to adhere to a specification absolutely, except where a clause in the specification allows a certain amount of flexibility for experimental or research work, interpretation of data, estimation of costs or completion date or indeed, anything else *agreed* by contract.

When interpreting the specification a designer may recognise that a new concept, perhaps in the form of an unusual and attractive feature, will contribute to an improved product but it is important that the customer's written agreement is obtained when any change to a major design concept or organisational matter is proposed.

We can summarise by noting:

1. In the commercial world, unapproved deviations from an agreed specification can result in financial conflicts and legal proceedings.
2. In the educational context, unapproved deviations may lead to lower grades and disappointment.

A brief summary on the interpretation and writing of specifications is given in Data Sheet 7.

In modern industry it is rare for a designer to undertake a major project in isolation; such projects are more likely to be tackled by a design team.

The prototype machines shown in Plates 1.1A, B, 1.2A, B are examples of major projects. These were both designed by a team of engineers consisting of the author (responsible for the mechanical design work) and a group of designers, technician engineers (highly skilled engineers who are capable of manufacturing virtually every mechanical part of such machines) physicists and electronic engineers. A glance at these photographs conveys the complexity of the projects and it doesn't take a lot of imagination to conjecture the importance of communication between all those involved. In addition to the main design team there are many other important contributors to the project, both within the company (management, accountants, quality control experts, purchasing department staff and safety officers, etc.) and a similar group of people associated with the customer.

A group of people working together always present a few problems (human nature making its contribution!) but optimum success in design objectives depends on co-operation, coordination and communication between all members of a team in a manner one would expect from an aircrew or team of mountaineers, where there is maximum effort from everybody to ensure the success of a 'project'.

Communication and interactions between members of a design team are highly complex activities; there are, however, certain basic concepts that help with the degree

Plate 1.1A   Prototype machine for air sampling and chemical analysis

Plate 1.1B

Plate 1.2A   Prototype recorder

Plate 1.2B

of communication necessary for projects such as those shown in the photographs:

1. Try to attend every project planning meeting; if there are occasions where this is not possible:
   (a) inform the project leader in good time;
   (b) check the minutes of the last meeting for reminders on any action;
   (c) make available to the meeting notes, sketches or data on matters of immediate urgency.

It may be argued that these activities are an obvious design process activity; this is true to a large extent when the design team are working in close proximity, possibly in one department; however, with major projects different areas of expertise can occupy separate departments, buildings or indeed parts of the country. In these circumstances the activities outlined above are of much greater significance.

2. Try to be punctual with data promised to other team members (whose activities are often tempered by anticipation of its availability).
3. Listen carefully to suggestions from other team members; anybody associated with a project can have inspiration.

Returning to the photographs – it is obvious that some form of control must be exercised in handling the masses of information generated by complex projects; a simple technique for this is demonstrated in Chapter 2.

1.7   PROJECTS

There comes a time in every design course when students are given the opportunity to undertake projects of their own choice. It is at this time when most requests for project subjects are forthcoming. Many students are ready and enthusiastic with a favourite potential project (for example, a musical instrument or photographic equipment, etc.) while others are seeking subject areas to explore and from which projects that 'grip the imagination' (and thus provide strong motivation) may emerge. It is here that advice can be given on useful project sources.

There is an enormous number of industries from which potential projects can originate (and the Appendix may stimulate inspiration on projects associated with any of the industries listed). What follows is intended to make a contribution to the student's *frame of mind* in the business of seeking project subject material. These are all open-ended brief specifications which, by slight changes in the wording, can be used at different levels of a design course to suit time available for a project, availability of finance, quality of product and design experience, etc.

1.7.1   Medical

Medical establishments are a rich source of project subjects. Many students are aware of this but have difficulty in deciding whom to approach. It is advised that initial contact is made with physiotherapy and occupational therapy departments. It is the author's experience that the staff of these departments are not only willing to suggest project subjects, but also take an active interest in the design of devices and the necessary interaction between designer, patient, consultants and other medical staff. They also have many contacts with other hospital departments and are usually most helpful in passing on requests for projects.

Other sources to explore for medical projects are homes and hospitals for the disabled, where many people are in need of special equipment for education, therapy or to enjoy a better life style.

It is important however, that a medical source is given constant feedback on the progress of the project and every effort must be made to complete and test a prototype product. One medical project has already been discussed – the grip trainer – and the source from which that came was a children's hospital. Here is another:

SOURCE: A disabled person
PROJECT: Weighing machine

Many people who spend their lives confined to wheelchairs have a weight problem. They have to check their weight regularly, in their own homes and (preferably) unassisted. A requirement exists for a weighing device on which the wheelchair can be positioned (by hand drive at the wheels); it must also be independent of any external power source and be provided with a weight indicator that can be easily read without the need to lean forward.

For other examples of medical projects see Data Sheets 8–10.

1.7.2 Science

Many projects can be devised which are related to other studies from the curriculum of a course. These have two useful advantages; first they remain in the department as permanent visual aids and second the student acquires a much better feel for the subject than can be assimilated from the mathematics (or descriptive text) only. This statement will be better understood when reading the following specification:

TITLE: Torsional stiffness comparator
DESCRIPTION

This product relates to engineering science and is intended for use as a visual aid to demonstrate how torsional stiffness varies from one material to another and also between solid rods and tubes. The basic need is a portable instrument, designed to stand on a desk, which will demonstrate stiffness differences by the angular displacement of a pointer.

Data:

- The instrument is to be designed to accommodate rods and tubes which have a common outside diameter and length.
- Specimens will be mounted singly and must be firmly clamped at one end. The free end will then be held by a device which allows each specimen to be subjected to an identical torque.
- The angle of twist will be displayed in a manner suitable for demonstration to a class of students.
- The angular displacement must operate a mechanical link to the display pointer – therefore it will be necessary to magnify the angular displacement; design calcula-

tions would establish an optimum torque value and length for a solid steel rod to
ensure adequate movement of the pointer.
■ An outside diameter of 10 mm is recommended for the test specimens.

For other examples of this type of project see Data Sheets 11–13.

1.7.3   Manufacturing Industry

A visit to a manufacturing company, with a request for projects which do not have a
short time priority associated with them often prove fruitful; here is one example:

SOURCE: **An engineering company**
PROJECT: **Stacking machine**

A large power press produces steel components resembling 400 mm × 300 mm trays.
The press is angled to a degree that allows ejected 'trays' to fall away from the press
tool and emerge at the back of the press. The project specification required a machine
which will receive ejected trays and – under mechanical control – stack them in batches
of 100.

Another example is given in Data Sheet 14.

1.7.4   Useful design workshop equipment

Time, space and equipment are always at a premium in design studio workshops. The
various types of models and prototypes produced there are of 'infinite' variety and most
of the manufacturing work can be achieved on standard types of machinery. There are
however, a number of tasks that occur repeatedly and cause delays and frustration.
These are often special, small quantity jobs which need to be subcontracted (if funds
are available) with the inevitable delays that small quantity subcontracting can bring.
Similar delays can occur through the need to make outside purchases for components
which can be adapted as a compromise.

Some of these 'delays' can be avoided by using prototype machines which have been
designed as a course project. The following specification explains one such project.

Other examples are to be found in Data Sheets 15 to 20:

TITLE: **Gear-cutting machine**
DESCRIPTION

A requirement exists for a special-purpose gear-cutting machine for the design studio
workshop. There are many projects and exercises in mechanism design in which gears
are used, but in many instances those made to a high degree of precision are not essential. Precision gears have to be ordered (often with an undesirably long delivery period)
and are very expensive.

Where the forces on the gears are light and minor backlash is tolerable, simple gears manufactured with a tooth shape closely approximating the involute often suffice.

A simple method of manufacturing this type of gear is to cut the teeth with home-made special milling cutters manufactured with a near involute profile and spacing the teeth positions with simple index plates similar to those used on an indexing head.

Design Data

Milling cutters and indexing plates can be manufactured in the college workshop.

Gear materials:	Duralumin and plastics.
Gear sizes:	maximum diameter 60 mm
	maximum gear thickness 10 mm
Gear types:	plain spur gears.
	spur gears with fixing collars.
	bevel gears.

Bench mounted machine.

Cutting action feed to be operated by hand.

Assignment

- Design and manufacture a prototype machine.
- Provide an instruction manual and a display board containing a representative set of gears.

General

Finally – look around at products which already exist and consider a project which specifies a re-design for a novel feature. Examples of this type are given in Data Sheets 21 and 22.

1.8   ADDENDUM

The next two Chapters are aspects of case studies undertaken for various industries; I have chosen these to demonstrate the value of certain techniques and to introduce other useful design concepts. The reader should bear in mind that the mental activities explained in Chapter 1 were always contributing to the ideas presented.

2

Project planning – a generator of ideas

A project-planning exercise is a process that is used mainly at the beginning of a new project; it helps to ensure a thorough examination of a specification and to coordinate ideas and information contributed by members of a design team and the people with whom they communicate. It also helps with analysis of conflicting requirements, design priorities, matters of administration and the integration of information from all sources. We have to devise ways of handling all this information, and if in so doing we can generate useful ideas, we produce a very effective management technique.

A simple analogy helps to explain the advantages to be gained from some form of information-handling system. We are all familiar with card games in which a player is dealt thirteen cards! Before playing the hand the cards are put into suits and each suit is ordered in card value. This is done because a player can then see, and appreciate the implications of, the many 'choices of scheme' available. As the game progresses, modifications to the original scheme, or confirmation of the choice of scheme, are made easier because the constant supply of information provided by other players cards can be easily assessed when referring to an organised hand; and, together with the experience of the player, new ideas are created, positive and confident decisions are made and the future course of the chosen scheme constantly assessed. Note – this all happens because the information available to the player is properly managed. Also note that the player is working exactly to specification – the rules of the game – which can only be changed in consultation and agreement with the other players. A great deal more than thirteen factors need to be managed with a design project and the outcome must *always* be a 'winning hand'.

Now let us examine the parameters which determine the need for an organised and careful management of a design project.

1. There is a need for thorough and detailed examination of design specifications regardless of the form in which they are presented.
2. A relationship exists between all parameters influencing design decisions. The acquisition of maximum information about a project, at the very beginning, helps establish a 'mental network' of these relationships; this is itself a generator of ideas.

3. There are advantages in living with the project and getting the feel of the job before major work on the drawing board is undertaken.
4. Early communication between everybody concerned with the project is essential. This ensures that all are aware of problems and priorities at the outset.
5. Careful project planning results in better scheduling, budgetary control, design organisation and cost estimation (particularly the early rough estimates).
6. Projects do not always progress smoothly; there can be delays and interruptions for many reasons (in some cases for many weeks or months). Project planning provides a useful method of storing information for later reference.
7. It is necessary to devise a simple method of identifying viable schemes from the many possible design options.

Practising designers will be aware that the elements of a project-planning exercise are not easily defined; there is a complex web of activity, often involving a whole team of people whose studies overlap, or are conducted in parallel, and where a bright idea from one member of a team can change the whole perspective of a project. However, if such activities are analysed we can identify elements of the overall process:

- project planning
- identifying viable schemes
- reviewing cost estimates
- selecting a scheme
- revising cost estimates
- producing formal drawings
- revising cost estimates.

Separating such activities from what is usually a more integrated process in practice, provides a useful basis for understanding the first step – project planning.

The subtle manner in which these links are interconnected to form the overall web of design planning activity can be described as *experience*, beginning with the subjects undertaken in a design course and enhanced by project work, industrial training and working in a commercial design environment. Such experience develops an ability to maintain a *constant visualisation of the whole web structure*; influences on the overall structure by vibrations occurring in any of its links, can then be recognised and assessed immediately they occur. Developing competence in this skill should be one of the main objectives of students' training in all design activities.

Design ideas will be generated in the mind of an experienced designer immediately on reading a specification. These will not be confined to mechanical features but to the whole spectrum of design activities including matters of organisation and administration. The overall direction of a designer's thinking and efficacy of application to the project will depend on many human attributes which can be described collectively as *ability*. These include breadth of experience, practical skills, academic training, drive, enthusiasm, enjoyment derived from designing etc. Since the development of many such attributes is the purpose of design training, students are advised to practise a disciplined and commercial approach to design projects at every opportunity.

Every designer has an individual approach to project management but, within the enormous complexity of their thought processes, there is common ground on which the foundation of good design practice is built; the common ground is used in this book to construct guidelines for a technique which, with a little practice, students will find simple and convenient to use. The objectives of this can be described as follows:

- thoroughness in the search for information
- constant focusing of parameters towards an optimum solution
- identifying factors which dictate compromises in design concepts
- constant communication with everybody concerned with a project
- frequent appraisal of economic factors and priorities
- persistency in the generation and examination of ideas.

An engineering case study follows to demonstrate a project-planning technique which encourages the mental disciplines necessary to achieve these objectives; students of other design professions can benefit from reading the study and to experience the technique, even though some of the language may be unfamiliar. The details are of greatest interest to the engineering designer but the technique is an information-handling exercise and generator of ideas – which is of interest to all types of designer.

A project planning exercise consists of a study in four parts:

1. specification analysis
2. review of priorities
3. study of basic concepts
4. summary of conclusions.

These are explained in detail as the case study proceeds (also see Data Sheet 23).

The study is a sequence of relatively simple procedures which (in Section 2.1) are somewhat disguised by the need to use many words of explanation. This situation is obviated in Section 2.2, which is a similar study, written without words of explanation.

2.1   CASE STUDY: ULTRASONIC SCANNING MACHINE

2.1.1   Specification

Title:	Ultrasonic scanning machine
Customer:	Quality control department of an aluminium alloy manufacturer
Product:	Extruded aluminium alloy bars
	Length 2–6 m
	Width 100–150 mm
	Thickness 25 mm

Description of process
- The machine is required for automatic inspection of bars by ultrasonic flaw detection technique (a *flaw* is a defect in a material – a void, area of porosity or an inclusion).

- Bars will be placed in a water-filled tank and scanned longitudinally by an ultrasonic transducer.
- The effective diameter of the transmitting crystal is 20 mm This will be positioned approximately 2 mm from the surface of the product under test, thus maintaining a constant water-filled gap between the crystal and the product.

  For convenience the product and transducer will be referred to as specimen and probe.
- The pattern of scanning will be line-overlap; the probe will move laterally 12 mm for each longitudinal scan.

In operation, ultrasonic pulses are transmitted by the probe through the water coupling and into the specimen. Reflections of the pulse are received at the probe from the top and bottom surfaces of the specimen as a constant signal. If the specimen contains a flaw or foreign particle an additional reflection is detected between the top and bottom reflections.

Thus, when displayed on an oscilloscope the time base is set to display the top surface pulse on the left-hand side of the screen and the bottom surface on the right-hand side; any signal appearing between these two 'peaks' indicates a flaw (see Figure 2.1).

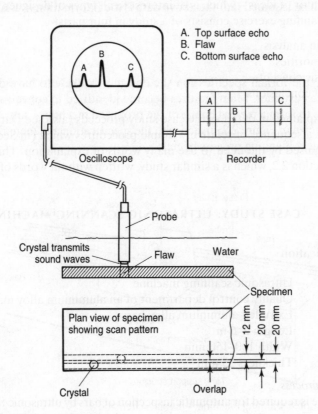

A. Top surface echo
B. Flaw
C. Bottom surface echo

Figure 2.1   Ultrasonic scanning process

Methods of recording
- Signals from the probe will be recorded on heat-sensitive paper – providing a separate record for each specimen.
- A recording device must move the heat stylus in direct proportion to probe movement.

Special features
- Scanning speed variable, 8–32 m/min
- Provision must be made for independent manual control of probe movement to facilitate inspection of individual flaws, using an oscilloscope display.
- Gap width between probe and specimen must be variable between 0.5 mm and 4.0 mm.
- probe dimensions: length 200 mm
  diameter 25 mm
  cable length 2 m

Additional to the above would probably be a request for cost and completion date estimates.

In practice the specification is likely to be more detailed, but for the purpose of this case study it will be assumed that the above data is the information available when the project starts.

Notes

Before reading the formal specification the designer is likely to have discussed the project with the customer. Main features become identified in conversation and the designer begins to form a mental image of the proposed machine – something similar to the sketch shown in Figure 2.2; armed with the specification and mental images such as this, the Project Planning process begins:

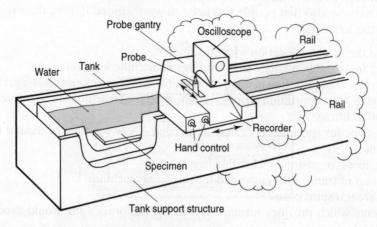

Figure 2.2   Ultrasonic scanning machine – mental schematic

2.1.2 Specification Analysis

This activity is conveniently discussed in three phases:

1. Project review
2. Survey of design parameters
3. Specification study.

Project Review

The project review involves a careful examination of the project objectives. It costs very little in time or effort to survey a specification with a view to generating such questions as:

- Is this the best way to solve the problem?
- Are we likely to infringe a patent?
- Has the customer given enough general information?
- Why was the contract placed with us?
- Can we utilise anything within our past experience?
- Where shall we build it?
- How would it be moved?
  etc.

Notes

Quite a few pertinent questions can be generated by such discussions between two or three members of the design team, at the very beginning of a project; experience shows that occasionally a factor arises which can have far-reaching consequences on the management or design features of the product. Consider for example the last two points, location of the proposed machine and its possible transportation from the place of manufacture to its final location. Bearing in mind that the machine will be long (over 6 m) and narrow, also that people will need to work around it, then the two queries generate the following:

- Discuss the machine location with customer.
- Inspect floor and immediate surroundings at machine location area.
- Design with transportation in mind or, assemble on site – seek customer's preference.
- If design for transportation is preferred the machine structure will be influenced by method of lifting.
- Floor space for specimen bars may be needed – how will they be loaded into the machine?
- Check access to customer's loading bay.
- What sort of transport is envisaged for this large machine?
- What are transport costs?
- A scheme which provides for assembly at the customer's site would avoid many problems!

Notice that important design parameters begin to be considered even at this early stage of the study.

Note also that all the other points listed here (and others that may arise from a general discussion about the main objectives of a project) will first contribute further ideas and queries, second accelerate the process of getting the feel of a design venture containing unfamiliar features, and third begin to orientate the designers thoughts towards concepts which emerge from an overview of all considerations – getting the project off on the right track from the very beginning.

The reader will also notice, from the few typical factors listed, an immediate integration of technical and management parameters. This occurs throughout a design study and forms the foundation on which the web of knowledge about a project is mentally structured.

One other important point: the project begins with a formal project review exercise as described above, but an overview is kept on a project at all times – the process never ends and is constantly in mind in all that follows.

Survey of Design Parameters

This involves a careful reading of a specification – underlining and noting everything that may influence the design, together with *potentially useful ideas* that emerge, as follows:

General observations
- Recorder must switch on with start of scan.
- A scan and recorder stop signal will be necessary.

Specimen bars
- Will surface protection be required when loading (lifting points must avoid specimen support points in tank)?
- Specimens could be loaded by hand (weight 60 kg); discuss with customer.

Water tank
- Maintaining water lever – will a mechanism be required?
- A chemical inhibitor may be necessary – algae, corrosion etc?

Probe
- Will water turbulence be a problem?
- Probe immersion depth?
- *Note*: accuracy of probe carriage will depend on level of rails.
- Some means of locating probe crystal centre will have to be devised.
- Obtain a probe quickly – check best method of attaching to it.

Scan pattern
- Possibly adjustable limit switches for both directions of probe travel.

Ultrasonic flaw detector
- Would need to be positioned on probe carriage (for individual flaw inspection and limited cable length).

- Obtain information on weight and size etc.
- Protection of leads (mechanical and from water).
- Stability and vibration problems?

Recording
- Coding of record chart and specimen will be necessary.
- Safety interlock to check recorder – possibly a fixed test plate (containing holes) at one end location of specimen, would provide a simple check on every chart and at every 12 mm scan.
- If recorder is fixed to probe carriage a simple gearbox would ensure accurate scan/record chart ratio; obtain information on heat-sensitive paper, control unit and stylus.

Gap width (between specimen and probe crystal)
- How accurate? And what tolerance is acceptable (relates to carriage rail accuracy)?

Notes

The data given here is all suggested by a close scrutiny of the specification, and ideas provoked by each feature have been listed (including those where expertise obviously plays its part: but the important point to note is that a formal survey of design parameters provides the motivation for a careful study of the specification while raising new queries and contributing useful ideas).

Specification Study

Although several design queries have already been noted, the specification is once again scrutinised and, for each *main element*, the question is posed: 'is there anything we do not know concerning ...? Answers are also sought to any queries generated by this question, and to any other queries not yet settled:

Automatic inspection
- The maximum scanning time is only about 10 minutes – since specimens and charts have to be prepared and loaded between scans there will probably be an operator present at all times – this may influence design features – discuss with customer.

Specimen bars
- To what degree are specimens twisted or curved?
- Will specimens be inspected in batches of one size?
- Are specimen lengths infinitely variable?
- Can the machine be designed for a range of discreet specimen sizes?
- How are the specimens to be removed from the machine after inspection?

Water tank
- Is there a preferred working height?
- What plumbing services are planned by the customer?
- Probe carriage will need to clear ends of specimen – is there any limit to space available at customer's factory?

Scan pattern
■ Would there be any advantage in scanning two bars at a time – using two probes and two chart recording heads in parallel? (Flaws on either specimen could be individually inspected simply by changing a cable connection to the oscilloscope.)

Ultrasonic flaw detector
■ Electricity supply? It may be necessary to design a 'concertina' (curtain) cable system – alternatively, an overhead cable support high enough to accommodate full carriage movement (can we buy a system?).

Recording
■ Dust – will the machine be used in a clean environment?

Gap width
■ How accurately must the probe crystal face be set parallel to the specimen surface?

Notes

Thus we have made a concentrated effort to study every piece of information, and every resulting idea, that can be extracted from the specification; and, in so doing, have raised many queries. In practice the answers to many of these would have been obtained quickly, in conversation with customers, equipment suppliers and colleagues etc., with some answers promoting a few more queries. Before proceeding to the next stage of the study it is useful to list answers to the queries raised so far; bringing these together underlines the value of the specification analysis exercise and helps with discussion on the next phases of the study:

■ Load specimens sideways into tank from a table parallel to, and 2 m from, the machine.
■ Load by overhead lifting crane – 'harness' will be provided by customer – get details!
■ Possibly manufacture tank in three sections – fitted together at site – plastic sprayed for waterproofing before fitting gantry rails (address of spraying company …) – (get quote for spraying).
■ Gantry rails would have to be in sections (eliminating transport or access problems).
■ Scan stop signal to be a large, prominent red light.
■ Start scan at same position for all specimens – interlock with start switch.
■ Scan stop to be set for each size specimen bar by a simple adjustable limit switch trip.
■ Specimen surface protection – use a softer material for specimen supports.
■ The position of specimen supports must relate to loading harness supports (to clear each other on loading!).
■ Probe case will be plastic material.
■ Probe immersion 150 mm (approx – not critical).
■ Position the flaw detector on probe carriage (short probe leads are required).
■ Flaw detector weight … dimensions as catalogue.
■ Recording charts will be individually numbered. Specimens will be coded by a stamping operation (on ends) when located in lifting harness (to minimise errors).

- Enthusiastic response to test plate suggestion.
- Maintain gap width at 2 mm ± 0.25 mm for most purposes.
- An operator will attend the machine at all testing sessions.
- Specimens will not be twisted significantly but occasionally curvature could be 2 mm (this means that probe must 'ride' the specimen to maintain gap width).
- Specimens are to be inspected in size batches (generally) – occasionally the machine will be required to inspect a single specimen.
- Assume specimen lengths are infinitely variable within the specified limits.
- Specimens to be removed from tank by loading harness.
- Preferred working height (specimen surface) 800 mm from floor.
- Provide water input pipe and drain hole fittings (customer to specify).
- Space needed for probe carriage overrun is no problem.
- Other scan patterns have been considered by the customer and rejected – proceed as specified.
- Machine will be used in a clean environment.
- Probe to be mounted vertically – tolerance ±2°.

Notice that positive design features are beginning to emerge with this data; the contributions that a careful project planning study makes to the overall process of designing is obvious!

2.1.3 Review of Priorities

Following a specification analysis study, when a much greater feeling for the project has been acquired, there are advantages in highlighting and discussing priorities relating to the design and administration of a project.

Some priorities may have been discussed during earlier studies or in conversation with the customer, but these have such a large influence on decision making that a *concentrated* effort at this stage will help considerably with the overall management of a project. (Even a conversation on what constitutes a priority can pay dividends!)

Here then is a typical 'discussion' relating to our case study:

1. Perhaps we should investigate the frequency with which serious flaws are likely to occur in the specimens to be tested. If *infrequent*, the manual control facility may not be worth while; located flaws are identified on a recorder chart and could be further investigated by hand-held probes when the specimen is returned to the loading table. The gantry design would be simplified and costs reduced. Perhaps the customer hasn't thought of this – check with customer!
2. Another feature worth discussing with the customer is the proposed mechanically integrated recorder; is this vital? there are recorders available which would give a similar record: such a device would need to be mechanically linked to the gantry (for varying lengths of specimen) but it could be envisaged as a simple bought-in item (this could be cheaper!).
3. What work force are we anticipating for the design and manufacture of this product;

and how will this project relate to existing and anticipated contracts within our own company?

4. How realistic is the required completion date; is there any flexibility?
5. Are we negotiating a maintenance contract? If not, we must discuss spare components and repairs etc.!
6. Has anybody discussed the design 'life' of this machine?
7. Is there any likelihood of a repeat order for this, or a similar machine, from the customer or anybody else?
8. What about an operator's instruction manual? this will involve time and cost!
9. Will we install and commission the machine?

There will be other priorities, but this list is sufficient to explain how priorities appertaining to a project are explored; it also provides a means of demonstrating how the outcome of such discussions further influence the management of a project. To this end the following comments are numbered to equate with those in the above list:

1. The specified manual control feature is to be retained.
2. Similarly for the specified recording feature.
3. Our workforce is somewhat stretched! Design with subcontracting in mind (possibly the tank assembly, supporting structure and cable system).
4. The completion date is not flexible; include a contingency figure in the costing review, for any extra effort which may be needed later!
5. The customer intends to maintain the machine! We will quote for spares in the future if required!
6. Testing of these specimens will continue into the foreseeable future: long-term corrosion problems must be considered and due consideration given to the electrical supply cable system (to avoid fatigue problems with wiring connected to the moving gantry).
7. Repeat orders for similar machines are a distinct possibility: Keep this in mind at the detail design stage!
8. An instruction manual will be prepared by the technical sales department in parallel with our own advertising literature – no costing problem!
9. We will install and commission, show this as a separate item on the revised costing sheet!

Notes

Notice how decisions on any of these factors will have a marked influence on subsequent thoughts and design ideas. It is also worth noting that all types of project will involve major administrative factors and that occasionally a review of priorities is useful in highlighting *conflicting* priorities; hence promoting concentrated discussion on such issues.

After completing the first two phases of a project planning study (Specification Analysis and Review of Priorities) it may be necessary to write a supplement to the original specification (or write a revised version). This must then be formally agreed with the customer.

2.1.4 Study of Basic Concepts

Until this point in the study thoughts about the project have been influenced by the processes explained in Chapter 1 under the heading Problem Solving, involving (1) some free thinking (2) concentrating on major subject areas and (3) thinking of the product as a closed system; details, design features and matters of administration have now been studied and visualisation of the main features has been established. These are:

1. a water tank into which specimens will be placed for ultrasonic scanning located in a structure which will also support rails along which the gantry will travel;
2. a gantry structure housing an ultrasonic transducer (probe), recorder and oscillo-scope, which moves longitudinally along the rail track.

Sufficient details about these features have been considered to enable the intro-duction of another useful process which is intended to encourage concentrated thought about *each main feature*. The procedure involves:

(a) *summarising* what has been provisionally decided and giving some thought to *design implications*;
(b) consideration of parameters which *link the major features*;
(c) *identifying queries and information sources* (which help in focusing design options).

At this stage, however, we will allow our thoughts about each feature to include consideration of the following:

- *unconventional solutions*: (new patents are in this category);
- *fundamental principles*: mathematic, scientific or know-how;
- *new technology*: materials, processes, techniques;
- *visualisation*: of the product in operation;
- *inputs* to the product: human, machine or environmental;
- *problems*: anything potentially detrimental to the life or operational efficiency of the product.

Thinking about a project in these terms helps to extend design concepts beyond those which leap immediately into the mind; the subsequent comments about details are followed by a brief note on the design implications resulting from some of these influ-ences and generally from this phase of the study. The details are listed neatly for the sake of clarity but, in practice, it is inevitable that concentrated thought about one feature triggers ideas about others; the designer is scanning the whole project constantly and inspiration about about any detail can occur at any time.

Tank and Support Structure

Item: tank dimensions
              length 6 m plus overrun at each end to clear space for loading specimens
              width 150 mm plus clearance each side for probe carriage (and possibly the
              lifting harness)
Design: the overrun allowance must accommodate the complete length of the gantry
              front apron (a new idea arising while *visualising* the gantry at one end of
              the tank as a specimen is being loaded).

Item: water level
Comment: 150 mm approximately
              no mechanism required (top up occasionally)
Design: check with customer that there is no possibility of ground subsidence due
              to the weight of water (a new thought promoted by *problems* that might be
              associated with a large volume of water).

Item: test plate
Comment: customer to provide details
Design: to be in a fixed position at one end of tank and clear of specimen space.

Item: tank structure to be fixed
Comment: agreed with customer
Design: keep in mind that this may be subcontracted.

Item: specimens to be 800 mm from floor level
Comment: specimen support pads must not cause damage to specimens
Design: possibly use hard rubber or plastic pads positioned to accommodate lifting
              harness components.

Item: main structure might be conveniently constructed in three sections: or *use
              a brick-built structure* lined with concrete and sprayed with a plastic lining
Comment: this second option was conceived while thinking of *unconventional* solutions
              such as swimming baths, ponds, horse troughs etc., which all contain water.
              The customer was contacted and was immediately keen on the idea
Design: a brick-built structure would avoid all problems of sealing, leaks and corro-
              sion: anchor pads for mounting specimen support pads and rails can be posi-
              tioned by a wooden jig during construction
Comment: this idea is *adopted* as the best of all options, by the designer and customer,
              and will influence the overall design from this point in the study
Comment: the reader is urged to notice where in the planning study this important
              design decision occurred and what brought it into mind; the word *tank* had
              put a biased thought into the designers mind i.e. tank interpreted as *metal*
              tank!

Item: rails – possibly use bright mild steel square section material
Comment: could use standard lengths placed end to end

Design: these would need to be plated (possibly dull chrome) to protect against rust level by graded packing at rail support pads

Comment: level specimen support pads by gauging from rail pads after these have been levelled (new idea).

Item: water inlet pipe and drain

Comment: details from customer

Design: ensure that this information is available for tank structure design and inform the plastic spraying company of details before finalising the design (another thought about potential *problems*).

Queries

Item: tank structure – get quotation for brickwork

Comment: find plastic spraying company

Design: discuss the tank proposal with spray company before finalising the design.

Item: brick structure

Comment: is any external finish required? discuss with customer (a new idea brought about by *visualising* the finished structure).

Item: chemical inhibitor

Comment: get details from supplier

Design: check that an inhibitor is compatible with all components that will be immersed – specimens, specimen supports, harness, probe test plate and plastic tank lining material (this thought was provoked by considering *problems* that may result from using a chemical inhibitor in the water).

Probe Gantry

Item: gap between transducer crystal and specimen surface to be 0.5 –4.0 mm

Comment: a support wheel will need to be designed to ride slightly curved specimens to maintain a constant gap

Design: the wheel will need to be large enough in diameter to protect the probe if the end of a specimen curves upward by up to 2 mm (a *new idea* which occurred while visualising the probe reaching the end of a specimen and reversing for another longitudinal scan) visualising this function also generated the concept that the probe must sit between *two* wheels of the same diameter, so that the crystal can cover the specimen completely (right up to the edges of the specimen)

Comment: discussions about the relationship between probe and specimen established that the scan area could exclude 3 mm along both edges of the specimens (see Figure 2.1).

Item: probe support and fixing to gantry
Comment: the probe will move up and down as its support wheel follows curvature of
              specimens
Design: support near to the lower end and towards the top, at the gantry, for smooth
              action (*fundamental principle* relating to supporting a long shaft).

Item: specimen lengths 2–6 m
Comment: plus an allowance for test plate length
Design: limit switches, adjustable over scan lengths, to be associated with gantry
              support rail.

Item: scan to move transversely in steps of 12 mm

Item: manual control of gantry
Comment: control must be provided in both longitudinal and transverse directions
Design: some form of clutch will be necessary for changing from automatic to hand
              control.

Item: probe length 200 mm, diameter 25 mm, cable length 2 m.

Item: scan start position
Comment: will be the same for all specimens
Design: location indicator for start position might be associated with a corner loca-
              tion nest for all specimens (a new idea which occurred while *visualising* the
              placing of a specimen into the tank)
              possibly an indicator light related to gantry and probe holder position on
              rails (longitudinal and transverse): this is a *new idea* emerging from thoughts
              about the start position.

Item: scan stop signal
Comment: to be a red signal light placed prominently on the gantry.

Item: probe immersion
Comment: to be approximately 150 mm

Item: probe vertical tolerance ±2°
Comment: mechanical adjustment will not be necessary – crystals are always mounted
              in probes to a closer tolerance.

Item: environment
Comment: considered clean enough for normal operation of machines.

Item: probe wheels
Comment: the axle pins and wheels will operate under water
Design: use stainless steel for pin and the correct grade of tufnol for wheels (check
              with supplier).

Item: limit switches
Comment: position at each end of travel to prevent overrun when using manual control

Design: possibly a lever type microswitch and alarm buzzer: also provide for transverse movement of probe-holding platform.

Item: handwheels for manual control of gantry
Comment: discuss type
Design: place at an ergonomically acceptable height if possible; if too high (because of gantry design demands) consider providing the operator with a platform (a *new idea* while *visualising* the operator using the manual controls).

Item: flaw detector
Comment: to be firmly fixed to gantry
Design: discuss with supplier.

Item: electrical supply to gantry
Comment: use a cable curtain to avoid loading problems
Design: suspend behind tank structure and connect to back end of gantry: get quote from supplier.

Item: recorder chart
Comment: a new idea which emerged from *visualising* a chart being fitted concerns possible *problems* that might occur due to mistakes in changing the charts – particularly following an ultrasonically clean specimen.
Design: a safety interlock signal associated with a chart-retaining clamp would prevent the machine from functioning unless the chart had been changed (discuss this *input* function with customer).

Item: flaw detector
Comment: water may splash this electronic device (thinking about matters which may be *detrimental* to this instrument)
Design: design a cover which will not cause overheating of the instrument: talk to supplier.

Item: specimen length/recorder chart ratio
Comment: the recorder stylus moves with the probe so that chart length represents specimen length; a change of ratio might have advantages when scanning short specimens
Design: investigate the possibility of incorporating a variable gearbox between gantry travel and output movement to recorder's stylus (a new idea brought about by *visualising* the appearance of a chart after scanning a short specimen).

Item: manual operation of gantry
Comment: check degree of control required
Design: some form of gearing may be necessary to achieve the degree of sensitivity necessary (a thought generated by the basic concept *input*).

2.1.5   Summary of Conclusions

What has been listed here is only a fraction of the mental activities associated with the project; in practice every experience would have been focused on the study, including of course, much more that is associated with mechanisms, calculations and control system; so many thoughts and ideas that recording them in detail is out of the question! It has been necessary to record details here to explain the designer's thinking when influenced by the *basic concepts* approach. It is useful however, to make a record of final decisions and data emerging from the study and to make this available to everybody associated with the project. An example of a chart written for the features discussed in this case study would look something like Figures 2.3 and 2.4 (*Note*: these would normally occupy one large sheet of paper.) A summary chart does not contain every minor detail, because these have now been firmly established in the mind; it serves

Tank structure	Gantry	Gantry	Manual control
Brick-built line with concrete plastic spray inside (subcontract)	Scan/record ratio to be variable (design a mechanism)	Longitudinal scan speed variable 8–32 m/min	Bar type handwheels at 800 mm from floor
Rails 20 mm square section bright mild steel (plated)	Gap under probe 0.5–4.0 mm setting by eye against a scale is adequate	Rail lengths to accommodate full gantry width when clear of specimen	Handling sensitivity to be 1 revolution of handwheel for 10 mm probe movement
Level rails with graded packing	Probe to ride specimen	Gantry travel limit switches at rails adjustable for various specimen lengths: the switch at the 'start' end can be fixed (*new idea*)	Design an engage/disengage clutch
Provide fixed test plate (customer drg. ref. ____)	Transverse scan 12 mm	HARNESS	SAFETY
Water inlet and drain (details to subcontractor)	Safety interlock to avoid overrun Location indicator for crystal centre	Ensure clearance between specimen supports and harness slings (drg. no. ____)	Provide an emergency stop button (*new idea*) Large red light to indicate completion of a scan (customer request)
Chemical inhibitor (supplier ...)	Power supply by cable curtain (supplier ...)	Check the above for *all* specimen lengths (*new idea*)	

Figure 2.3   Summary of conclusions – example 2.1 chart

Specimen	Recorder	Flaw detector	General ideas
Lengths 2–6 m	Heat-sensitive paper type – persue the possibility of buying and adapting a suitable instrument	Fix by existing screw holes in base (remove feet)	Use three point support for gantry (thinking about *fundamental principles* for stability
Width 100–150 mm		Weight ...	
Thickness 25 mm	Design interlock for scan/recorder *on* together	Dimensions ...	Guide wheels each side of front rail will ensure transverse stability (*new idea*)
Curvature 2 mm maximum	Design interlock to ensure chart change at each scan	Design splash screen (avoid cooling louvres)	Design inching mechanism for transverse scan in 12 mm increments
Specimen supports in plastic or hard rubber material: provide with method of level adjustment (gauge from rails): could be by graded packing if supports were suitably 'nested' in tank base (*new idea*)	Design a mechanism to vary scan/recorder chart length ratio	PROBE Depth of immersion approx. 150 mm Provide easy access for probe changing Design support wheels unit for maintaining gap	Contact Quality Control departments (ours and customers) to invite any other useful comments

Figure 2.4   Example 2.1 chart continued

as a useful reminder (particularly when delays occur or when working on more than one project) and is often a further source of ideas as the mind once more focuses on the design features.

Note that design ideas are still emerging as data is being written into the chart.

In practice it would be advisable, at this stage, to consider writing an updated specification and to agree this with the customer (or course supervisor).

After the new specification is agreed we would proceed with the first layout drawing armed with detailed information on every major feature of the project, in full consultation with management and customer and with an agreed interpretation of the specification.

Whatever the type of project, the criteria listed on pages 30, 31 and 32 need to be considered before a great deal of work is done on the drawing board, computer or in the design studio. A relatively small amount of time spent on a careful project-planning study will pay the sort of dividends demonstrated by this case study.

For major projects it is also an excellent method of 'briefing' everybody involved with a project while serving as a focus for discussions by the design team.

The study will also help in highlighting (if not already obvious) where computer work, materials testing or other research is needed and where a model should be investigated – which is the subject of Chapter 3.

A sequence of activities subsequent to a project-planning study is suggested in Data Sheet 24 – method of working and reporting.

Note: it is inevitable that words of explanation about a designer's thoughts and their influences on design details, conveys an impression that a Project Planning study is unduly complicated; to counter this impression the next case study is written without detailed comments:

2.2   CASE STUDY: CASTING MACHINE

This example outlines a study for another special-purpose machine; it provides the reader with an overview using the same project-planning techniques as for Example 2.1, but presented without explanatory notes.

2.2.1   Brief Specification

Description:	Machine for casting small brass ingots
Production rate:	60 per hour
Approximate ingot size:	Plan dimensions 150 mm × 40 mm
	Thickness 30 mm
	Corner radii 4 mm
	Taper on all sides 20°
Location:	To be installed adjacent to crucible supplying molten brass via a tap valve.
Casting method:	Direct into open moulds.

2.2.2   Specification analysis

Points Arising

1. Mould material must withstand temperatures of 1000°C.
2. Investigate thermal shock on mould material.
3. Will quenching of ingots be necessary?
4. Study method of ejecting ingots (if necessary).
5. Possibly invert moulds to eject ingots.
6. Identification mark cast into ingots?
7. Basic types of machine:
    rotating turntable (horizontal)

rotating turntable (vertical)
'conveyor belt' type
arm rotation type (with just a few moulds)
indexed slide type.

8. Effects of heat on mould conveying mechanisms.
9. Safety interlock – ensure mould ejection to avoid overspill.
10. Periodic mould dressing may be necessary. (automatic or hand?)
11. Mould unit cooling facilities may be necessary.
12. Prototype date?
13. Floor space available?
14. Cost target?
15. To function with or without an operator?
16. Investigate dripping at metal feed tap.
17. Cooling rate may influence production rate.
18. Variations in molten metal temperatures could be contained by variable speed control to delivery demand.
19. Necessity for metal feed chute? Might need a heater.
20. Investigate life of moulds – may need to provide for easy replacement fitting.
21. Maintenance will be by skilled engineer.
22. Control panel function to be established.
23. A heating unit between feed tap and the machine would facilitate pouring temperature control?
24. Check on standard components which might contribute to design features.
25. Query machine operating periods.
26. Purity of ingots – is mould contamination a problem?
27. Fume extraction – is there existing plant available that might be adapted?
28. Get safety and fire regulations for the machine location factory.
29. Review all corrosion possibilities.
30. Investigate lubrication requirements including heat considerations.
31. Investigate existing resources oil, air, power factor etc.
32. Possibility of pre-heating moulds if rapid cooling is a problem.
33. Check location (floor etc.) to design method of installation.
34. Query – is there a similar machine that could be adapted (or part of a machine)?
35. Production rate 60 per hour?
36. Check regulations and Factories Acts relating to metal casting.

Queries Answered to Date

6. Customer's trade mark to be indented in base of castings.
12. One year from date of order.
13. Floor space area 4 m × 4 m.
14. Feasibility Study £___
    Prototype £___
    Prototype Development costs to be negotiated.

15. Design for operation by unskilled labour.
23. Useful idea – temperature control at crucible not possible.
25. Virtually full time – night shifts not planned.
26. No serious problems envisaged.
27. Main duct close enough to be used if necessary.
34. Turntable indexing unit – (manufactured by _____)
    might be adapted.
35. Ingots to be stored – problems on production rate not expected.

2.2.3   Review of Priorities

1. Completion date for prototype and production rate (linked to market priorities).
2. Floor space : cannot be exceeded without minor structural changes to building: consider only if priority 1 is dependent – unlikely!
3. Prototype cost: rough estimates agreed: revise on completion of design study: make sure revised costs ensure priority 1 : customer agrees to flexibility on development costs.

2.2.4   Study of Basic Concepts

Note: This example demonstrates the use of charts for Basic Concept studies (Figures 2.5 and 2.6). These are more convenient than listing the details (as in the previous example) since a designer is repeatedly considering interrelationships between all the main features of a project and a chart provides a useful overview.

Note also that urgent queries are highlighted at the top of each feature column.

2.2.5   Summary of Conclusions

This summary is likewise given in the form of a chart (Figure 2.7).

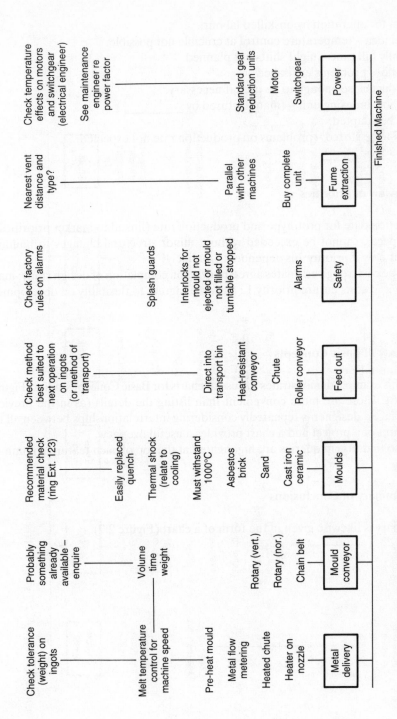

Figure 2.5   Summary of basic concepts – Example 2.2 chart

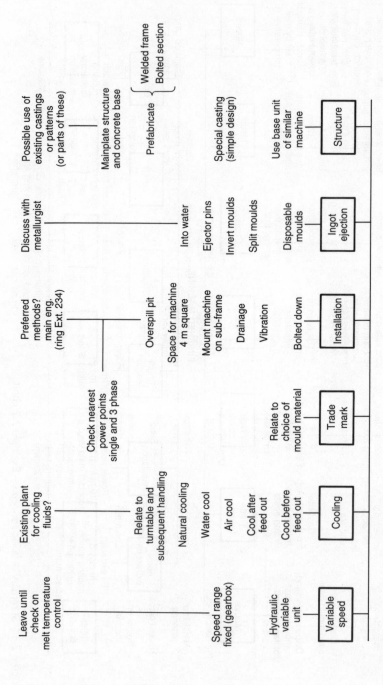

Figure 2.6 Example 2.2 chart continued

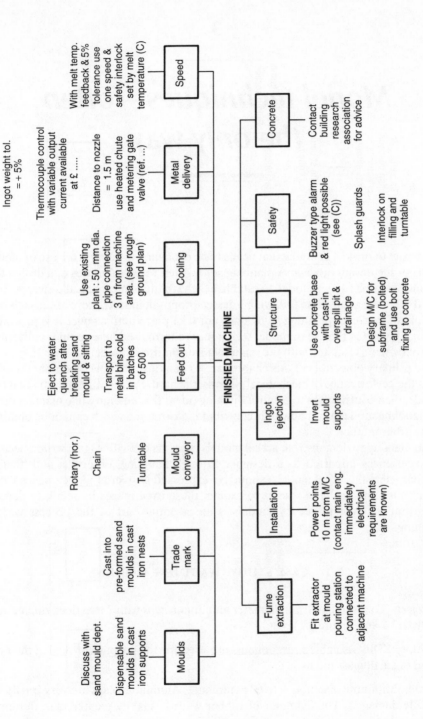

Figure 2.7 Summary of Conclusions – Example 2.2 chart

3

Model techniques – often the only way

One aspect of project planning that needs to be explained in detail is the use of models as part of the design process. A planning study often reveals features of a design that demand research using a model, to establish the viability of mechanical concepts. Such a need can arise at any time following a design proposal; indeed, there are cases where the bulk of a project-planning study does not take place until a concept is proved by research involving the use of model techniques (a term used here to refer collectively to all activities associated with the manufacture and use of models).

The primary objectives of this chapter are to demonstrate how models are used to assess the performance of mechanical systems, the nature of the valuable data that they provide, their contribution to creative thinking (ideas that emerge while manufacturing and handling models) and how they contribute information which cannot be obtained by any other method.

The inability to demonstrate actual models presents difficulties to a writer, because the experiences imparted to a designer through handling, observing and 'feeling' the characteristics of a model cannot be expressed in words alone; we can only compromise and use case studies to explain these experiences in detail; to demonstrate that the role played by models is an essential part of the process we call designing.

3.1 CASE STUDY: WADDING MACHINE

Project: To assemble 8 mm diameter aluminium *caps* with 1 mm thick rubber *wads* (Plate 3.1 and Figure 3.1)

History: The assembled components are used for the purpose of sealing the open end of small glass phials.

Data: Minimum assembly rate 60 per minute. Aluminium caps are very fragile and easily damaged. The diameter of rubber wads is slightly greater than the inside diameter of caps so that an airtight seal is formed on assembly.

Plate 3.1 Wadding machine – caps and wads before assembly

3.1.1 Solution

The first step towards visualising a solution is to play with the components (a process which usually pays dividends in the form of ideas when designing machines which handle or assemble components). A few caps and wads are twisted, dropped, squashed, rolled

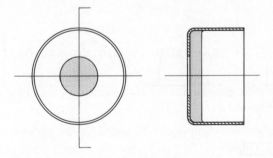

Assembled components

Figure 3.1 Wadding machine – assembled components

and probed with a small screwdriver. Squeezing the wads between thumb and forefinger revealed a useful stiffness before bending and this led to a basic principle of assembly where wads might be used to extract caps from a hopper in the sequence shown from right to left in Figure 3.2.

Model

Operation of the model is shown as a sequence in Plate 3.2A–D.

Notice the perspex windows which allows viewing of the mechanism in operation; these can also be swivelled aside for modifications to the model or dealing with jamming of components during experiments. Notice also that a simple method of using the end of a coiled spring as a leaf spring is to attach the whole spring. The *slider* is hardwood running on softwood guides (quite adequate for a few thousand operations).

In operation the slider moves from the right hand side position. At it moves to the left the *pusher* contacts the bottom wad and causes this to extract a cap from the *cap hopper* (see Figure 3.3). The *cap-retaining spring* controls the cap as it rotates out of its hopper, keeping it in contact with the wad until it is located on top of the wad. The two components are 'clamped' together, as they move further to the left, by the *cap-holding spring*. The slider comes to rest when components are located above the *wadding pin*. At this point the next cap and wad are resting on the pusher.

As the slider moves to the right, on the return stroke, the wadding pin is raised by the *cam* and assembles the components (pushing the cap-holding spring against a fixed machine surface, thus providing the necessary reaction force).

As the cycle is repeated the wadding pin is lowered, under the influence of a spring, before the arrival of the next cap.

After assembly the components are moved towards the *ejection window* by action of the pusher in moving a line of caps towards the assembly position (see Plate 3.2 D).

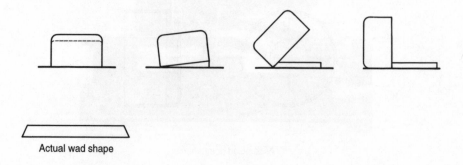

Actual wad shape

Figure 3.2   Wadding machine – sequence of assembly

Plate 3.2   Wadding machine – sequence of operation

C

D

Plate 3.2   (Continued)

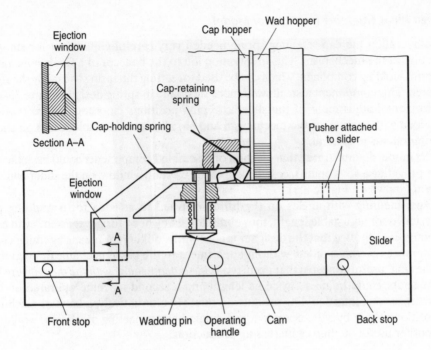

Figure 3.3   Wadding machine – machine function diagram

Features to be Investigated by the Model

It is here that reasons for using a model becomes apparent; questions which need an answer from the model and which *can only be supplied by the model*:

- Will repeated operations reveal any fundamental problems related to the assembly method?
- What are the best hopper chute clearances relative to (a) the tolerance band applied to component dimensions and (b) the assembly action?
- Is there an optimum weight of components, in each hopper, for smooth assembly action? Note: extra weight can be placed in both hopper chutes to study this feature.
- What approximate forces need to be imparted to the components by the cap-retaining spring and the cap-holding spring?
- How does the speed of operation influence the assembly action? Will high-speed movement of the pusher cause wads to bend (due to the inertia of the system) and cause failure?
- Will it be necessary to provide a means of ejecting the assembled components?
- What is the best shape for the cap-retaining spring?
- The wads are pressed from rubber sheet during manufacture and this results in tapered edges. Can these be fed to the machine either way up?

Design Ideas Emerging from Use of Model

- In operation the cap-retaining spring needed very careful adjustment for smooth action of the mechanism. This observation led to the concept of a finely adjustable spring-holding component which can be used for setting the spring to exactly the right force. This component also allowed more flexibility in spring design (Figure 3.4).
- Horizontal adjustment of the slider stopping positions: another feature that (a) helped with 'tuning' of the mechanism and (b) allowed wider tolerances on slider components in general.
- The model demonstrated that ejection of assembled components could be achieved by providing a 45° sloping face, adjacent to the ejection window, on the slider component (shown on Figure 3.3).
- Experimenting with angles on the lifting cam led to an idea for a wadding pin designed for adjustable length, thus eliminating any need for great accuracy in cam dimensions (notice that the designer is constantly looking for means by which close tolerances can be avoided, without impairment to the basic mechanical function).
- The final useful concepts that occurred *while experimenting with the model* were (a) the slider could be made twice its length and a second machine operated on the return stroke, thus doubling production rate; (b) alternatively, a bank of machines positioned side by side, operated by a single bar connecting all sliders, would be another useful method of increasing production.

3.1.2   Conclusion

The model was used to solve all the problems described above and produced many hundreds of assembled components. What can be achieved using a rough model can, of course, be much more reliable with engineering precision; costly prototypes will be designed and manufactured with a confidence that is absent if the model phase has not

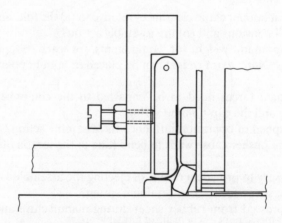

Figure 3.4   Wadding machine – cap-retaining spring adjustment

been attempted. Students should also note that the model costs very little if designed with care, whereas repeated modifications and remakes of prototypes can be very costly indeed. It is also necessary to emphasise that there is a limit to how roughly made a model can be, if anything useful is to be learned from it. As with this wadding machine example, all models have to be constructed in a manner that is rigid enough to impart engineering precision where this is needed.

This simple case study was chosen to represent the role of models in the design of machinery that handles and assembles components; not only related to the performance of the mechanisms but also in feedback of creative ideas to the designer. A similar exposition to this case study could be written about all such machinery. A designer does not always have the time or opportunity to manufacture a model, a situation which has its disadvantages, but the ability to supervise its design and an awareness of the features that must be incorporated, to ensure the required feedback information, is an enormous advantage; hence the importance of acquiring practical skills necessary to achieve this.

The next case study is chosen to demonstrate another important aspect of creative feedback; a mental process that has been responsible for countless advances in science and technology in every field of endeavour. This is known as serendipity – creative feedback brought about by events which are unexpected but which the skilled mind immediately recognises as a solution or having great potential (a 'eureka' moment).

3.2 CASE STUDY: TRANSFER MECHANISM

Project: To design a mechanism which is part of a special-purpose machine designed to dispense chemicals at a number of work stations.

A small glass dish is located on a circular turntable as shown in Figure 3.5. The

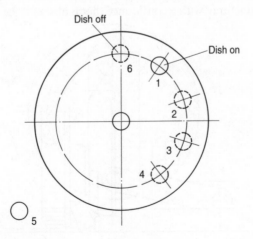

Figure 3.5 Transfer mechanism – schematic of turntable

turntable rotates and stops at positions 1 to 4 in turn. It is a feature of the machine that work station No. 5 needs to be placed at some distance from the turntable, where the dish must pause once more before being transported to position 6, where it is removed.

Data: time interval between work stations – 6 s: rest time at all work stations – 2 s: positional tolerance for dishes 0.5 mm (maximum) in any direction from specified centre points.

3.2.1   Solution

An outline solution was devised by visualising the placement of the dishes by a human forearm rotating about the elbow; the concept occurred while thinking about the problem of transporting a dish to position 5. This led to an initial idea of placing the dish on a pivoted arm which could be rotated when required to transport the dish to position 5 (see Figure 3.6).

It was then realised that the required arm movement could be achieved by shaping the arm so that it could be rotated, under the influence of the moving turntable, anti-clockwise by a suitably positioned block and clockwise by a wheel attached to a 'bridge' spanning the turntable (Figure 3.7).

It was also realised that the arm pivot needed to be provided with a slipping clutch, which is a device supplying enough friction to prevent the arm rotating freely; thus, it would rotate only under the influence of a force and remain in position when the force was removed.

Plate 3.3 shows a simple cardboard model of the mechanism which helped to

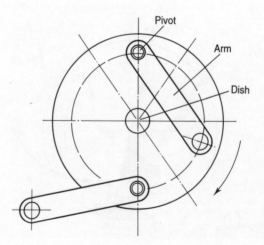

Figure 3.6   Transfer mechanism – initial concept

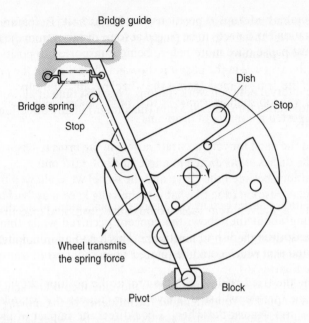

Figure 3.7 Transfer mechanism – bridge spring action

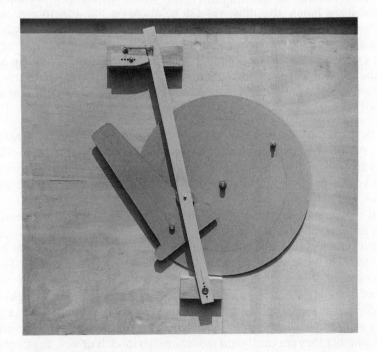

Plate 3.3 Transfer mechanism – cardboard model

establish the required mechanism geometry and Plates 3.4A, B a plywood model made
at a later stage to test the mechanism (the provision of two arms was decided at some
point in the project-planning study).

It was when the arm was in the position shown in Plate 3.3 that the point of this case
study occurred; the arm jammed against the wheel because the curved edge in contact
with the wheel was too shallow. (*note*: at that time the bridge holding the wheel was
fixed.) This suggested a number of problems:

1. how to avoid the need for very accurate curves in the arms to ensure accurate positioning of the dishes at position 6 to a tolerance of ±0.5 mm;
2. if slight clearance between the curve and the wheel were allowed the dishes could
   vary in angular position (due to what is technically known as 'backlash');
3. in the long term, uneven wear rates would cause positional inaccuracies.

To explain how all of these problems were overcome by serendipity, it is necessary
to describe, in the first person, what happened then:

I was holding the fixed bridge, with the arm at the position shown in Plate 3.3,
pondering the problems listed above while repeatedly jamming and un-
jamming the arm against the wheel. I could feel the impact of the arm while
holding the top end of the bridge (I could feel it slightly bending). This action
generated a sudden thought – *why not spring load the bridge at the top end*! I
realised immediately that this was the solution; as the arm rotated around
the wheel, driven by the moving turntable, a point would be reached when the
bridge spring, acting through the wheel, would impart an extra rotating force
(torque) to the arm until it reached a stop. If part of the curve was still in contact
with the wheel at this point, further movement of the turntable would only result
in a slight stretching of the bridge spring while continuing to press the arm against
its stop!

Thus the final position of the dishes would be independent of any great accuracy
in the curve profiles or wear rates; the mechanism would also be backlash-free.
The only dimension that needs to be reasonably accurate is the distance between
the dish centre position and the edge of the arm which contacts the stops (see
Plate 3.4A).

Note: it is of interest that the plywood model performed with perfect accuracy even
though it could be described as a 'rough' model.

In spite of this lengthy description the complete cycle of this mechanism is difficult
to visualise; hence the value of models (new designers would benefit from making a version of this mechanism to experience the advantages). But the whole
purpose of including it here is to demonstrate how a simple event, brought about by
the *handling of a model*, provided the perfect solution. Experienced engineering
designers will confirm that such events happen frequently when experimenting with
models – whether they are cardboard models, rough models or well-engineered prototypes.

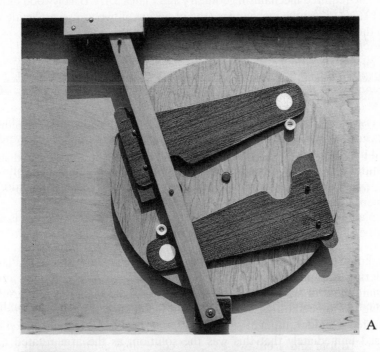

Plate 3.4   Transfer mechanism – visual aid

3.2.2   Conclusion

Students might usefully experiment with other methods of designing this mechanism, by plotting loci of the moving parts or by using a computer; the 'penalties' of using alternative techniques are (1) they need an enormous amount of time (2) are very complex and (3) as this case study has demonstrated, there is a possible loss of creative feedback.

It is a fascinating thought that a model created by a designer imparts valuable creative thoughts into the mind of its creator, including ideas which would probably never have occurred but for the making, handling and study of models.

Incidentally, the solution of this mechanism problem reminds us of another golden rule of design: endeavour to reduce the need for accuracy in manufacture without impairment to function.

3.3   CASE STUDY: FOLDING MACHINE

Note: This case study typifies those machines which process materials and for which a simultaneous study of mechanisms, their affects on a material and the quality of the process, is necessary.

Project: To fold and 'iron' two edges of corrugated cardboard strip as shown in Figure 3.8.

History: This project was undertaken by Rhoden Partners Ltd a firm of Design and Development Engineers in West London. Their brief was for a special purpose machine which would fold both edges of cardboard strip so that it could be rolled around cylindrical condensers as packaging. The folded edges providing end support as shown in Figure 3.9.

Data:
- Strip width before folding 140 mm
- Strip width after folding 108 mm
- Machine designed for fast and continuous production

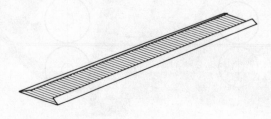

Figure 3.8   Folding machine – folded cardboard strip

Rolled corrugated cardboard

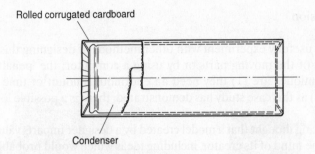

Condenser

Figure 3.9   Folding machine – wrapped condenser

- Strip material is fed continuously and cut to length (by another machine) after the folding operation.
- Edges to be folded towards corrugations (which run across the strip).

3.3.1   Solution

The solution to this project was dominated by the need to keep the cardboard strip taut while it passes through the machine. This was achieved by pinch rollers placed at each end of the machine (Figure 3.10) which have a common drive and speed of rotation – but the leading rollers are larger in diameter. The leading rollers are therefore trying to pull the strip faster than the trailing rollers will allow. Thus there is constant slipping at the leading rollers and the strip is kept in constant tension: an elegant and simple solution!

Model

The model used for prototype trials is shown in Plates 3.5A, B and diagrammatically in Figure 3.11.

Leading rollers

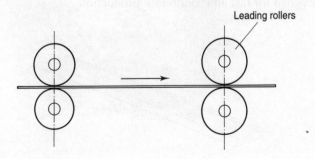

Figure 3.10   Folding machine – schematic of rollers

A

B

Plate 3.5 Folding machine

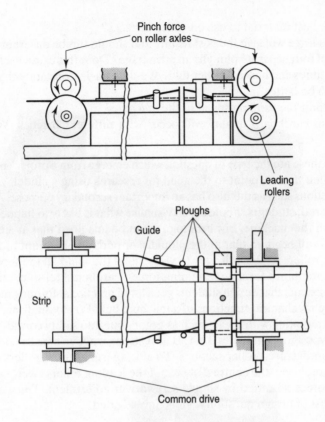

Figure 3.11   Folding machine – machine function schematic

In operation chain sprockets of equal size are fitted to the roller drive spindles. The cardboard *strip* passes through the first set of rollers and beneath a *guide* plate. Adjacent to the guide are a number of *ploughs* which progressively wrap the cardboard around the guide in the form of two returned edges. On leaving the guide plate the folds pass between the *leading rollers* which iron the fold in addition to providing constant tension and drive to the strip.

Features to be Investigated by the Model

The purposes of some models are best presented in the form of questions about the machine design which cannot be answered except by research with the model; this emphasises the need for the model and also for the particular form of construction needed if the model is to provide useful information:

- The leading rollers are larger in diameter than the trailing rollers. How much larger for optimum performance at high speed?

- What is the best material to use on roller surfaces?
- What pinch force will give the best result? and should this be different at each end?
- Is the use of four separate ploughs an advantage? Does this cause useful rupture of cardboard fibres and hence a better fold? Would a single, constantly changing profile type plough be better?
- What will limit the running speed of the machine?
- Variation in cardboard quality will occur with different batches. Will this be a problem?

Notice how these are the type of question which emerge from a project-planning exercise and, by their nature, point to the need for research using a model.

Posing questions such as this also has an important secondary purpose. As each question is considered, students should try to visualise what is likely to happen as the strip passes through the machine. For example, it can be imagined that at a certain speed the cardboard will begin to tear at the ploughs, or perhaps an abrupt tuck in the strip will cause the rollers to slip. Also try to visualise in the mind all the forces acting on the product – the pinch forces, the folding forces, cardboard tension, machine vibration etc. It is essential that design students get a lot of practice at forming mental images, not only of the mechanical functions of a machine but also at visualising other design parameters integrating with these, such as ergonomic and safety considerations.

The prototype shown in Plates 3.5A, B is of strong wooden construction to support accurately manufactured roller sections. To aid experiments the rollers can be easily removed and replaced, the centre distance of the leading rollers easily varied and the roller pinch forces modified by simple screwdriver adjustment. Thus, all the points raised in the list of design questions can be investigated.

3.3.2 Summary

Two other main lessons can be extracted from this example of model techniques:

The overall design of the prototype must be flexible and easily modified because each of the forces mentioned previously will be influenced, to some extent, by all of the others. Change one and the effect on others will be noticed. This may call for other modifications; the designer will use the model to 'tune' the forces for optimum performance and make a few mental notes of those aspects which ought to remain adjustable in the final version of the machine.

The design problems to be investigated by the model are not complete mysteries to some types of designer. Those with experience in paper-handling machines will know what is likely to give the best results. We have been discussing a relatively simple machine, but for more ambitious projects of this type one must consider carefully the possibility of calling in a consultant specialist. This can make good economic sense and save a lot of time; the designer must always consider every available option. The same comments would apply to other 'special' design work involving such things as optics, hydraulics, pneumatics, ultrasonics and virtually every project where the use of a new technology is proposed. The student of design must be aware that it is not necessary to design everything; part of a designer's skill is knowing when to use consultants and when to use bought-in equipment.

3.4 CONCLUSION

The case studies in this chapter have provided enough details and insights into the influence of model techniques on designing, to affirm that they have a fundamental, essential and unique role in the design process; and that all activities relating to the design and construction of models, including the requisition of the necessary practical skills, are of paramount importance – summarised by the following:

- *Creativity*: models provide a positive input of creative ideas, some of which result from observing the model performing its mechanical functions.
- *Research*: the word 'research' is defined as 'endeavour to discover factors by investigation'; a model is often the best possible research apparatus.
- *Economics*: for many design studies problem solving by a model is the easiest and cheapest technique to employ.
- *Serendipity*: models constantly present the mind with new and stimulating ideas; virtually every practising designer has experienced what may be described as 'eureka' moments while working with models.
- *Communication*: when any form of drawing is highly complex, models are often preferred for discussions between designers and customers.
- *Ergonomics and aesthetics*: the questions 'how does it feel?' and 'does it look right?', for many projects, can only be discussed with reference to a model.
- *Problem solving*: it is not always appreciated that some design problems cannot be solved by any other method.

It is also important that a designer is aware of the limitations to using models:

- *Dynamics*: only properly constructed prototypes will provide useful information on dynamic properties of a machine such as fatigue, vibration, resonances, noise etc.
- *Kinematics*: feedback data will be limited where temporary bearings, slideways, motor couplings, framework structures etc. have been used; components or assemblies will not move precisely in the manner planned for the engineered product.
- *Ergonomics*: models which are not specially designed for investigating ergonomic data will not provide useful information on the feel or sensitivity of levers, handwheels, pedals etc. or any other situation involving the handling of a product by an operator.
- *Aesthetics*: a model designed purely for functional purposes will not convey the aesthetic qualities of a product; the support of a good illustration or computer-generated image is worth considering for communication with customers.
- *Forces*: unless a model is specially designed as a research tool, information on the *long-term* effects of forces cannot be properly assessed; this includes stresses on components, backlash effects, impact forces, thermal effects on materials and virtually all other types of force to which a product is subjected.

There are many other aspects of modelling which are outside the scope of this book; a number of relevant subject areas are listed in Data Sheet 25 as a reference to further reading. Periodic literature surveys on these topics has significant educational value and new designers would benefit by using design library facilities rather than one or two specific texts, thus gradually building experience by becoming familiar with the whole spectrum of model techniques used in engineering and product design.

3.5   CONSTRUCTING DESIGN MODELS

This final section is an introduction to some of the practical methods used in the construction of models of the type described in Chapter 3. Included are methods of obtaining the degree of accuracy necessary for assessment of mechanical function and a number of devices useful for this type of model construction.

That models play an essential role in project planning has been demonstrated; this section will provide a foundation on which the necessary skills may be developed.

With the passing of time and the inevitable accumulation of experience, students who are developing as engineering and/or product designers will learn that the range of techniques and types of design model are virtually limitless. They vary from simple cardboard and drawing pin models – perhaps to investigate simple linkage mechanisms or escapements etc. – to prototypes made with engineering precision and almost indistinguishable from the intended final product. There are also those models which are hybrids to every degree, including those models which are made with adequate precision but which have a relatively short 'life' – enough to explore (say) a geometric relationship or other short-term research requirement; the wadding machine described earlier falls into this category.

Each type of model (except near-perfect prototypes) involves the utilisation of whatever materials are available for manufacturing the 'rough' part of the model and to build in the degree of precision demanded by the product function under investigation. The case studies discussed earlier demonstrate the need to develop the ability to do this quickly and economically; this section is therefore intended to assist with the process of developing these skills.

The techniques used by designers in manufacturing design models are too numerous to describe here (many are required for special products which involve unique features and research studies). However, the following notes are intended to provide a *foundation* on which the thinking processes, associated particularly with the manufacture of design models, can be established – and ultimately expanded by experience:

First, some general notes: when fitting components onto a wooden (or any other type) of 'rough' framework, a few basic concepts must be kept in mind:

- A framework will not provide for accurate location or assembly of components with engineering precision.
- The 'life' of a rough model is limited and it is therefore (usually) uneconomic to over

design it; screws rather than well-made joints and nuts and bolts rather than special fittings, are adequate for many purposes.

- There will be much handling, modification and adjustments during model trials. Remove all burrs and sharp edges, always use smooth wood or metal and make sure that any exposed points on woodscrews are filed flush with component surfaces.
- Always provide adjustments to one model support ('foot') so that stability in operation can be achieved on any surface. This is very important with large models where the weight of assemblies can cause gradual warping of a wooden base or local distortion of a fabricated metal structure.
- The framework or structure must absorb any forces imparted by functional requirements and must therefore be robust enough to withstand these without distortion. Avoid the possibility of bending or twisting the main support 'base' (a common source of unwanted problems). Well-seasoned planed wood or thick plywood make good bases (which can be easily strengthened by base runners if necessary).
- Think carefully about which parts of the model need easy access (and from which direction) and which parts need to be adjustable – so that the optimum performance of the model can be achieved.
- Think carefully about the 'core' of the device – those parts that must function with engineering precision – and then visualise the type of 'rough' structure needed to support this without influencing its function by distortion.
- The 'key' to assembling a prototype model is the accurate location and alignment of the functional components. To achieve this the designer must choose one of the accurate components as a *reference datum*. It is also convenient to provide a reasonably straight edge, to the base on which the model is to be mounted, as a location reference for components to be assembled very close to their final position (with reasonable model materials this may be closer than 0.5 mm).

A well-used method of final location and alignment depends on the use of over-sized holes and packing (also described as shimming).

Imagine that the first step in constructing a model is the need to mount the rod shown at A in Figure 3.12 vertically – with engineering precision – and to provide for minor adjustment in the horizontal plane.

Imagine also that the bracket holding rod A has been made from a close grained hardwood and crudely screwed together. This means that rod A will not be vertical – within the order of precision required – when the assembly rests on its base.

To bring rod A into the vertical plane with accuracy, use packing in the manner shown – first in one vertical plane and then in the vertical plane at 90° to this.

Notes
- It is advisable to make these adjustments separately, *lightly tightening the screws between measurements*.
- There are a range of thin metals (described as Shim) commercially available; a small stock of these is always useful in a design workshop. Useful sizes are 0.025 mm, 0.05 mm, 0.75 mm and 0.1 mm.
- It is also worth remembering a few other material thicknesses: cigarette paper 0.04

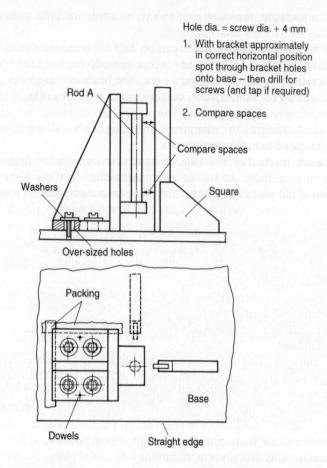

Hole dia. = screw dia. + 4 mm

1. With bracket approximately
   in correct horizontal position
   spot through bracket holes
   onto base – then drill for
   screws (and tap if required)

2. Compare spaces

Rod A

Compare spaces

Washers

Square

Over-sized holes

Packing

Base

Dowels

Straight edge

3. Insert packing and lightly
   tighten screws

4. Check vertical alignment of
   rod A and repeat 3 if
   necessary until degree of
   accuracy is acheived

5. Repeat 2, 3 and 4 for 90° plane

6. Release grip of screws for
   minor adjustment in the
   horizontal plane (2 mm approx.
   in any direction)

7. Lightly tighten screws for
   final check

8. Fully tighten screws and if
   required insert dowels

Figure 3.12   Vertical alignment of a rod

mm, newspaper 0.075 mm and typing paper 0.1 mm. Shim or other thin materials can be used in combination with each other or with thin gauged metals of greater thickness than shim, giving every thickness of packing likely to be needed.

■ To facilitate positioning of rod A in the horizontal plane, provide over-sized screw holes to allow for final adjustment.

- After final adjustment two dowels or roll pins can be inserted to assure absolute stability.
- In other circumstances – when rod A can be held in its accurate vertical plane by some other feature of a prototype – it is possible to measure the required thickness of packaging by assessing the gaps under the bracket using *feeler gauges*. If this method is used, always *make a final check* with packing in position and screws lightly tightened.
- This method of changing the alignment of components, *will not be suitable* if the packing is positioned close to the screws.

This is just one method of achieving adjustment to a precision component attached to a crude structure; there are various other techniques that could be employed, but the intention is the same – and this method is simple and utilises readily available materials.

Where parallel faces between two brackets are required, position one bracket accurately as described in Figure 3.12. Then clamp the second bracket face to the first as shown in Figure 3.13 using parallels or other suitable spacers – making sure that it is touching the base at one or more points.

Finally, measure spaces beneath the bracket – at screw positions – using feeler gauges. Insert the packing and tighten the fixing screws.

Check that parallelism between the vertical faces has been achieved as the spacing parallels are removed.

A similar technique using shim can be used when long (and relatively small cross-section) straight rails are required. In this situation the rail is first screwed to the base or supporting structure. A straight edge and feeler gauges are used to discern where packing is required (see Figure 3.14). Shims of the required thicknesses are then placed at each screw position – preferably in the form of made-up washers, or alternatively by providing two narrow shims placed either side of each screw.

When straightness accuracy is required in the horizontal plane, the *rail* is first drilled with dowel holes and the fixing screw holes in the base structure are drilled over-size

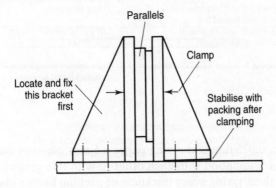

Figure 3.13   Parallelism between vertical faces

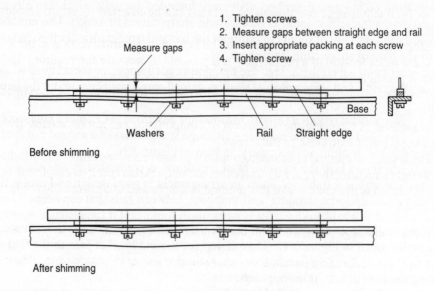

1. Tighten screws
2. Measure gaps between straight edge and rail
3. Insert appropriate packing at each screw
4. Tighten screw

Measure gaps

Base

Washers   Rail   Straight edge

Before shimming

After shimming

Figure 3.14   Rail straightness – vertical plane

(see Figure 3.15). Curves are then levered out of the rail at each fixing screw position (adequate straightness can be achieved by checking gaps between rail and a ground straight edge by using feeler gauges).

When the rail is straight – at the screw position being adjusted – tighten the screw.

This procedure is continued along the rail until the required degree of straightness is achieved.

Finally stabilise the rail by drilling the base for dowels – through the holes already drilled in the rail – and insert dowels.

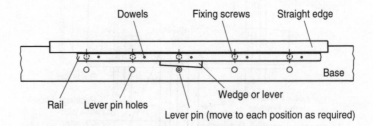

Dowels   Fixing screws   Straight edge

Base

Rail   Lever pin holes

Wedge or lever

Lever pin (move to each position as required)

1. Lightly tighten screws at (or near) touching points
2. Lever or wedge rail to close gaps – separately at each screw
3. Progressively tighten screws as each adjustment is made
4. Make a final check and then insert dowels

Figure 3.15   Rail straightness – horizontal plane

Note: Both techniques described above are intended for rails which are already 'straight' to commercial standards (i.e. without sharp curves or kinks). The methods explained will take care of slight curvature of the bar and irregularities in the straightness of the base or supporting framework.

One other point – it is advisable to use hexagon-headed screws or socket head screws for fixing the rails – and always provide washers or load-spreading plates under the screw heads.

Where two rails are required (as with the machine described in Chapter 2) the second rail can be levelled in the same manner – although straightness of a second rail, in the horizontal plane, is often not necessary, as seen in Figure 3.16.

The figure shows how the movement of a platform is controlled by *one rail only* – the other rail acts as a stabiliser. Such an arrangement is good design and avoids the problems associated with 'binding' and 'crabbing' between rails that can occur if both rails were used as wheel guides. This is particularly important if the distance between rails is relatively large compared with the distance between guide wheels.

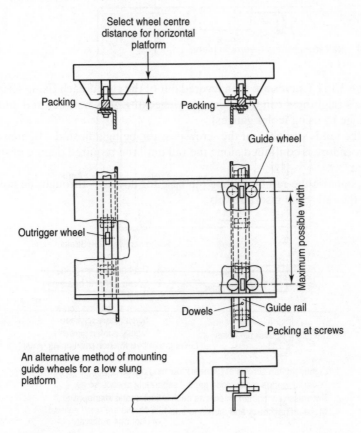

Figure 3.16   Platform support – two rails

One other point about moving platforms – keep the distance between the sets of guide wheels as large as is practicable for optimum stability.

The need to adjust the second rail to the exact vertical height of the first can be avoided by adjusting the height of the 'outrigger' wheel.

It is usually necessary to ensure that both rails are *parallel* in the horizontal plane. This is achieved by mounting the second rail lower than the first by a small amount. Each end of the second rail is than packed to equal levels *relative* to the first rail.

The procedure from then on is precisely as described for one rail – measure the packing required at each screw position.

Thus we provide perfect three-point stability and smooth trouble-free movement of the platform.

Note: With such an arrangement it is obvious that any driving mechanism must impart its forces to the platform at (or close to) the rail supporting the guide wheels.

Providing a stable three-point support is also important where a rod or tube is used as a rail. This case is shown in Figure 3.17.

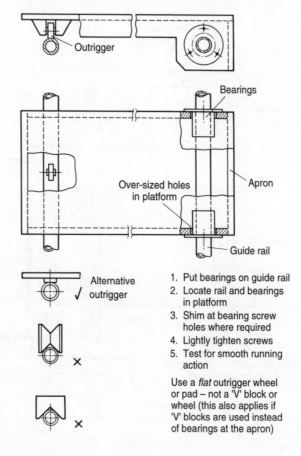

Figure 3.17   Platform support – two rods or tubes

Notice that the platform is provided with two bearings widely separated and that the 'outrigger' is again simply acting as a stabiliser.

One other useful application of using packing for achieving accurate and/or stable location on a 'rough' framework, is with the fitting of instruments or delicate apparatus. When these are placed on a framework it may be necessary to orientate or level them in relation to each other, or another part of the device being studied. It is also necessary to ensure that holding them to the framework by screws at fixing points does not cause distortion.

Both these situations can be managed by first locating accurately, measuring gaps by feeler gauges and providing suitable packing.

Before leaving the subject of accurate location of a component there are a few situations where a rod (in the form of a spindle, pivot or axle) is already orientated in space by another component of an assembly and needs to be provided with bearings (Figure 3.18).

It must be noted here that the base plate may not be perfectly flat and will not be absolutely parallel with the rod axis.

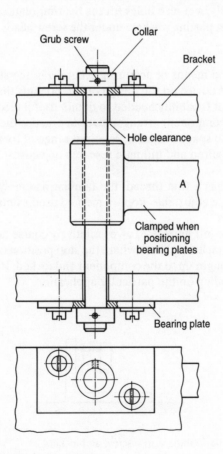

Figure 3.18   Spindle support – bearing plates

The solution to this problem is obtained by utilising two thin bearing plates in which holes have been machined which are slightly larger in diameter than the rod (say 0.1 mm).

The component (A) supporting the rod is first positioned at its best location relative to the base and firmly clamped by some convenient method. Over-sized holes are then drilled in the brackets at the correct axle height (the over-size diameter accommodating any slight angle between the rod axis and base).

The brackets are then fixed to the base.

Finally, the bearing plates are positioned at each end of the rod and fixed while located on the rod.

Since the bearing plates are relatively thin (say 3 mm) any slight deviation in squareness between rod axis and bearing plates is accommodated in the clearance between the bearing plate holes and rod diameter (shimming at the bearing plate holes will not be necessary).

Note: This method of location can be upset by the use of woodscrews at the bearing plates, causing the plate to move (or be pulled) out of position. The problem is avoided by using slightly over-size holes for the bearing-plate screws, careful centring of the screw pilot holes, placing washers under the screw heads and using good quality screws.

We have now studied means of acquiring accuracy in the structure of a prototype model. What follows is advice on the design of components that can be used to give a prototype the degree of flexibility needed to obtain useful feedback.

Note: these are just component arrangements; in practice the ideas must be adapted by due consideration to space available, required range of forces or spring pressures, degree of accuracy required and ultimate research objectives.

Using a screw which has a fine thread, the familiar assembly shown in Figure 3.19 provides a very accurate adjustable stop – a method used extensively for slider adjustment on machine tools.

The assembly shown in Figure 3.20 gives relatively coarse adjustment but is useful for some applications, including those where the stop position of a component is established and the stop brought up to the component and locked. It is used with or without the guide plate, depending on the particular application.

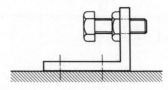

Figure 3.19   Adjustable positioning stop – screw and locknut

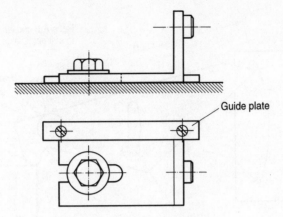

Figure 3.20   Adjustable positioning stop – slot and guide plate

The screw head and locknut, shown in Figure 3.21, provides a relatively coarse adjustment for a limited movement. Variations of this type of stop can be made by brazing any desired component shape to the screw head.

Where space is readily available, the lever type stop (Figure 3.22) can be made to

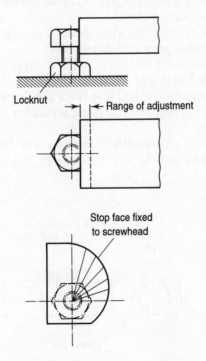

Figure 3.21   Adjustable positioning stop and shaped screw head

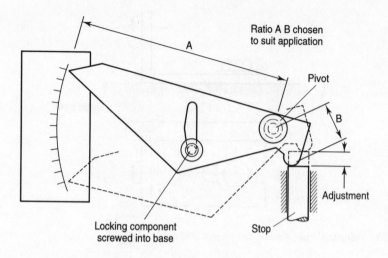

Figure 3.22   Adjustable positioning stop – lever type

provide great accuracy, with a useful degree of control on the stopping position, using a scale marked on a paper stick-on label.

Note: this technique works well only when the lever pivot is carefully made (without detectable clearance between pivot spindle and hole).

An eccentric disc (Figure 3.23) is a very useful and versatile method of obtaining an accurate stop adjustment. The degree of eccentricity can be chosen to suit virtually any situation. Rotating the disc by 180° will give twice the degree of eccentricity in stop adjustment. Again the pivot hole must be fitted accurately to its spindle. (A reamed hole is usually adequate.)

Note how small degrees of eccentricity provide very fine adjustment. When a stop position is established and the stop needs to be locked, a dowel or roll-pin is inserted.

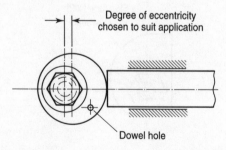

Figure 3.23   Adjustable positioning stop – eccentric disc

Another useful device is known as a slipping clutch (also familiar as torque-limiting mechanisms). They are also used where a component needs to be rotated to a position, held by friction in that position, and subsequently moved to another position (a cycle continuously repeated in intermittent motion mechanisms).

Figure 3.24 shows a simple arrangement where friction is provided by the clamping action – by washer or spring – at the centre of rotation.

Figure 3.25 shows a number of techniques which may be adequate for a particular application, but where variable control of the friction force is limited by the choice of components and minor adjustment to central locknuts.

The arrangement of Figure 3.26 provides much more control of the friction force, and when experiment is needed to establish the optimum force the spring is easily changed.

When larger friction forces are required, simple adjustable devices can be devised from whatever materials are available. These can be placed at two (or more) positions around the rotating 'table' (Figure 3.27).

A few more ideas which demonstrate how simple devices can be manufactured quickly and cheaply from materials readily available in the workshop are illustrated.

Figure 3.28 shows two arrangements for varying a buffer force.

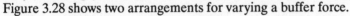

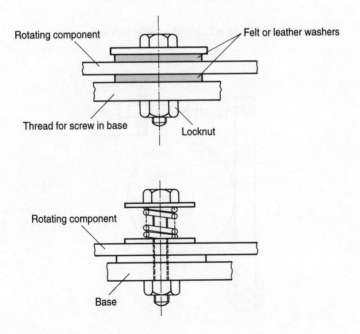

Figure 3.24   Slipping clutch – washers or spring

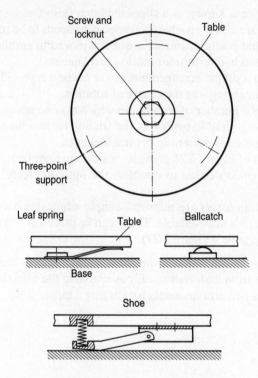

Figure 3.25   Slipping clutch – spring action with central locknut

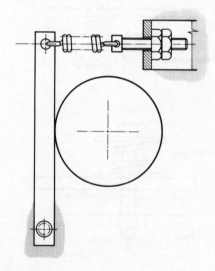

Figure 3.26   Slipping clutch – lever and spring

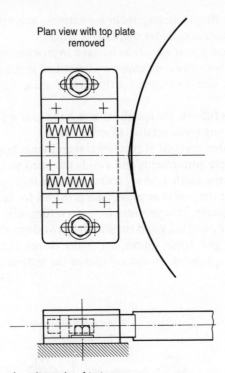

Plan view with top plate removed

Figure 3.27 Slipping clutch – disc edge friction

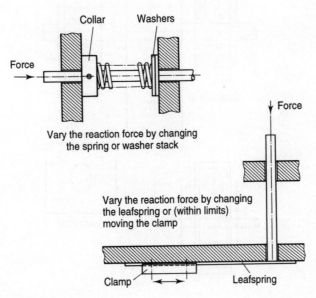

Collar

Washers

Force

Vary the reaction force by changing the spring or washer stack

Force

Vary the reaction force by changing the leafspring or (within limits) moving the clamp

Clamp

Leafspring

Figure 3.28 Adjustable buffers

Figure 3.29 shows a simple arrangement for variable axle force (this is the method used for the paper feeder and folder discussed earlier).

Figure 3.30 shows how a gear wheel can be used to provide accurate rotational increments to a spindle. As with all the mechanisms described in this chapter there are many variations of the basic idea. Examples for this device are as follows.

1. An additional disc is fitted to the spindle to which the gear is fitted, providing a useful method of zero setting (gear relative to spindle).
2. A stick-on paper label marked at rotational increments larger than a single tooth facilitates quick angle adjustments and avoids the need to count teeth repeatedly.
3. A slotted fixing for the tooth pawl is another method that provides for zero setting.
4. An eccentric pin, at the pawl bar pivot, also provides for zero setting (particularly for small toothed gears). This provides for zero setting with great accuracy.
5. *Note*: if the pawl bar pivot is a good fit, and the pawl shaped to a V which is slightly less acute than the gear tooth spaces, the device is free of backlash at the spindle.
6. All the above concepts relate to the spindle as the output component. By driving

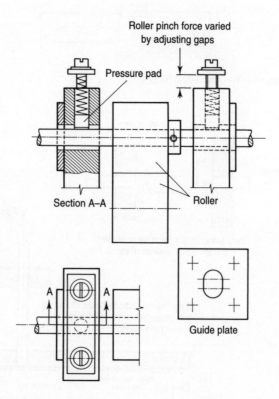

Figure 3.29   Variable axle force

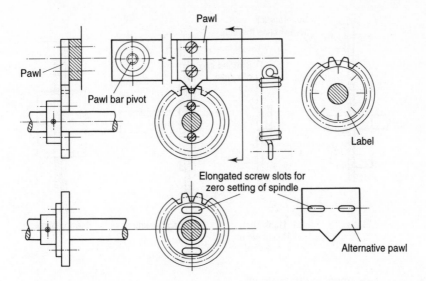

Figure 3.30   Rotational increments by gear wheel

the gear with a meshing gear, the pawl bar can be pulsed as an output (rotating about its pivot). This is only useful for slow speeds and the pawl must be made with a relatively shallow V and of a material which is suitable for rubbing across the gear tooth surfaces.

Obviously the particular application will 'design' the device – but these examples demonstrate that every device can be 'mentally examined' for versatility in potential application, thus helping to build a large reservoir of experience in mechanism design.

The use of thin materials – plywood, aluminium, hardboard and plastic sheet – can present problems of attachment. Simple holes in these materials are inadequate for many purposes, e.g. centre bearings, axle loading, weight supporting etc. Figure 3.31 shows four techniques used to overcome such problems.

Note: Buckling can often be minimised by pop-rivetting angle, U-channel or strip along edges (or, in the case of aluminium, putting a shallow bend along the edges).

Finally, some advice about the *components* used for prototype or experimental models. Time and money (and moments of frustration while queuing for machines or waiting for 'subcontractors') can often be saved by careful thought about the purpose of a component and its role in the model.

Where forces are low, or the designer is concerned only with mechanism geometry, demonstration or static models etc., components – or working surfaces on wooden replicas of components – which may be perfectly adequate can be fabricated quickly and economically from materials readily available in the design studio workshop. Three examples of fabricated components are given in Figure 3.32 and these will help to

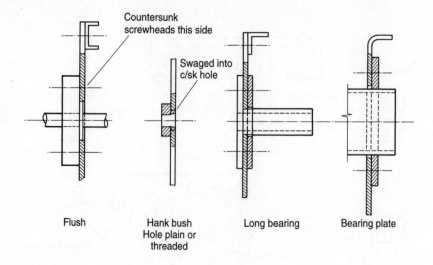

Figure 3.31   Attachment to thin materials

establish the concept of fabricated model components' as an important facet of model techniques.

Making a bobbin type component on a lathe is rather time consuming, particularly if it is large in size. Two plates swaged onto a shouldered tube (using a hammer and a steel ball) will suffice for many purposes.

A reasonable bearing can be made in a wooden component by inserting a piece of tube. To obtain a tight fit, first use a slightly undersized drill before drilling to tube size.

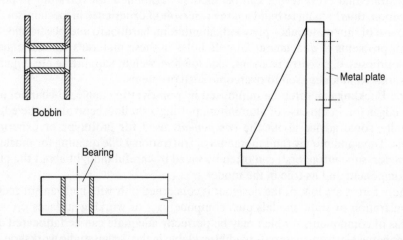

Figure 3.32   Examples of fabricated components

Waiting for components to be cast is often frustrating. Consider making a wooden near-replica and providing metal surfaces wherever required.

The reader will have already realised that the number of devices which can be contrived for prototype models is endless, and I have reached my limit for this book. There is just one other point I need to make now and this concerns the mounting of electric motors: It is important that motor shafts are *never* directly attached to a driven shaft. Always drive a machine using a belt or flexible coupling. The point to remember is that the motor shaft and driven shaft will never be exactly aligned; the coupling prevents strain on the motor shaft and bearings.

3.6   SUMMARY

Thus by discussing simple components we have seen how engineering precision can be achieved for experimental design models, mechanisms or 'rough' prototypes. In practice components and assemblies are often more complex. But means can usually be found to achieve accuracy by studying (1) how a component or assembly can be located and held in the correct accurate relationship to another, and (2) where packing, wedging, screw adjustments – or any other secure and reliable method – can be utilised to keep it there.

It is obvious that every prototype model is virtually unique and that just a few examples can be included here. The application of various techniques explained so far, begin to provide an insight to what was said earlier about providing a *foundation* on which the *thinking process* associated with prototype models can be established. I emphasise this point because a whole book devoted to the manufacture of prototype models would only constitute the 'tip of the iceberg' in prototype modelling technology; each model has to be individually designed and the designer's store of knowledge (coupled with natural ability, and ingenuity) tapped for the solutions to each design feature. This is another reason why a solid grounding in practical work is so vital in early training.

It is necessary for a designer to develop a 'feel' for this type of activity, and the only way of achieving this is to design and manufacture your own prototype models; a great deal of experience (and often valuable innovation) is lost if the building of prototype models is left to others (however skilled) particularly in early design studies.

Concluding remarks

A number of relatively simple case studies have been used to demonstrate various design techniques and to explain procedures essential to good design practice. These particular studies were chosen because they provide enough design factors (not technically or mathematically complex) to serve as a useful reference in early design studies. It is inevitable that students will develop their own preferred methods of handling information and project management as experience is gained (particularly in the final year of a design course and in industry); however, it has been my purpose to demonstrate essential basic principles which will assist with this development.

It has not been possible, in a book of this size, to include other types of project; but the basic design concepts which have been explained are applicable to them all. What will vary are the parameters influencing design decisions. For example, a mass-produced product will involve parameters related to activities such as company policy in marketing, long- and short-term effects on productivity and sales, target date and advertising, packaging, distribution of labour, utilisation of plant, etc. Notice that all of these activities have their experts (and often departments of specialists) but it is vitally necessary that the designer *communicates with them all,* takes into consideration their ideas, comments and priorities, has some formal method of handling and sifting all this information and ensuring that all members of the design team are fully informed. Notice also that comments emerging from any of these sources have some influence on the final product design.

Large projects need even more management and it is likely that the designer will develop procedures which 'link' the major assemblies and record results of mathematical analysis, scientific research and testing, etc. With large projects the design team is large in number and it is obvious that a formal project planning programme is vital. The designer is constantly in touch with other people in the company, e.g. works director, sales manager, consultants, quality control managers etc., all of whom may have differing ideas on priorities relating to a product being designed. Any method of visual communication (analysis results or model demonstrations) that assists liaison between all concerned with the product, is invaluable.

In a commercial situation there are many decisions influencing a project, which are outside the control of the designer. Most practising designers have experienced at some time the frustration of being told of a policy decision late on in a project which results in major design changes. It is inevitable that these situations will always be with us, but they can be minimised by ensuring that discussions related to essential decisions are dealt with early in a project.

I have defined *designing* as focusing the whole of one's experiences in life on the management and solution of a project. By appreciating and practising the design disciplines described in this book young designers will constantly 'sharpen the focus' of the experiences they will derive from all educational and industrial sources.

The book has been specially written for new designers; avoiding references to information and techniques that may not be available at an early phase of a design training programme. It has therefore been necessary to limit discussion on thought processes introduced in the case studies, without detracting from the primary objectives for which they were reported – as a practical guide for beginners. In practice the designers' mind will be occupied with many additional thoughts including mentally rejecting and accepting ideas, reasoning various courses of action and the likely consequences, focusing momentarily on parameters which may be peripheral to possible solutions while seeking inspiration – so many thoughts that any attempt to capture them all would be impossible. Indeed, there would be subtle influences too difficult to describe – such is the wonder of the human mind.

Appendix: Specialist design activities

Agricultural machinery
Aircraft industry
Architecture
Astronomical and optical
Bicycles
Building industry
Ceramics
Chemical industry
Civil engineering
Clothing and footwear industry
Communication industries
  (television, radio, telephone)
Defence industries
Domestic products
Electrical and electronic industry
Food-processing machines
Furniture industry
Gas and water industry
Illumination industry
Machine tools
Machinery for electronic
  components
Medical equipment

Mining industry
Motor cycle industry
Motor industry
Musical instrument industry
Office equipment
Oil industry
Packaging machines
Pharmaceutical industry
Plastics processing machines
Playground equipment
Precision instruments
Printing industry
Production machinery
Railway and transport
Research apparatus
Robotics
School equipment
Shipbuilding and marine
Structural engineering
Textile machines
Toy industry
Woodworking machinery

Data sheets

PACKAGING DESIGN

This data sheet was completed using the thought processes described in Section 1.2 (problem solving); it illustrates that application of the same processes to a specific area of interest results in a very thorough examination of the subject.

The procedure is simple:

(1) List everything you can think of (free thinking).
(2) Identify major feature associated with the subject (in this case handling, weatherproofing, transport, advertising, materials, costs, etc., etc.) and revise the list.
(3) Think of the subject as a closed system with input, output, activity and waste features.

Packaging is a word we use to describe the multitude of different ways in which we wrap up products for protection during storage and transportation.

The design of a package obviously depends on the type of product and the degree of protection desired by the manufacturer or retailer. There are also other (not so obvious) factors to be considered; but before discussing these we must examine the main basic functions of packaging in general:

- protecting the contents from contamination (e.g. food);
- protecting us from the contents (e.g. toxic substances);
- protecting the contents from physical damage;
- making it easy to handle and store products;
- protecting the contents from the weather;
- using the package to advertise;
- using the package for communicating instructions;
- utilising the package for a second purpose (e.g. certain camping and mountaineering products);
- enhancing the prestige and image of a product (e.g. expensive jewellery);
- conveying affection in the form of surprise gifts, birthday presents and special occasions.

As with all design activities packaging design requires a thorough study of its intended function. In addition to the precise protection properties (often studied scientifically), such as shape, size and resistance to various forces, etc. there are many other factors to be considered. Here are a few typical parameters which must be resolved before a package can be properly designed:

- Sales figures for each type of product showing sales trends for single items and orders for batches.
- Existing packaging costs and anticipated costs in the future if no re-design of packaging is made.
- Types of packaging available related to presentation, weight, strength (wet, dry

and crushing), weatherproofing, size, value of product to be contained in the package, cost of materials.

- Methods of sealing packages including labour costs, postage rates, regulations for transportation by rail, air, road or sea.
- Economics of producing compartments in the package for ancillary equipments such as connectors, leads, spare parts etc.
- Special design features for home and overseas markets particularly relating to storage and tropical requirements (heat, humidity, insects, dust, etc.).
- Import and export regulations at home and at overseas markets.
- Economics and maintenance considerations for returnable packages.
- Advertisement value in producing a package that the customer will find useful for other purposes: e.g. a suitable package becomes a permanent home for the product, or becomes a useful box, which keeps the manufacturer's name displayed.
- Value of the package if it is designed to be part of the production process for the product it contains. A package can often be a useful holding jig.
- Tolerable breakage rates.
- Methods of presenting company names and trademarks.
- Storage costs at home, during transportation and at overseas agencies.
- Cost of package production in batches related to anticipated sales, i.e. it may be cheaper to manufacture or purchase, say, three years' supply.
- Sales, advertisement and publicity value of package related to aesthetic considerations.
- Packaging regulations for all forms of transport.

Many people take packaging for granted – only concerned about the contents; this short study of the complexities of packaging design serves two purposes – to demonstrate (1) the need for careful and detailed study of packaging and its *total* objectives relative to a product and (2) that an enormous industrial effort is associated with packaging, involving designers, communicators, advertising experts, materials experts, graphic designers and artists, accountants, safety officers, inspectors and many types of scientist.

The range of packaging design is enormous – from a sweet wrapper to packaging space satellites – and the role of packaging in the peripheral work associated with project management must not be under-estimated.

DATA SHEET 2

PRACTICAL ENGINEERING TOOLS AND PROCESSES

Commonly used tools and processes used in design studios and workshops are
noted in the form of checklists. These are a useful reminder for obtaining a
working knowledge of any that are unfamiliar.

The headings should be interpreted broadly: thus, 'woodworking – hand tools',
implies becoming familiar with both the tools and their use.

The lists may not be complete for all areas of design interest, but students are
encouraged to extend them wherever necessary.

Cutting tools

Centre drills   Milling cutters
Drills   Negative rake
Form tools   Tool bits
Grinding wheels   Very small drills

Marking out and measuring

Angle plates   Parallels
Depth gauges   Slip gauges
Dial gauges   Square
Dial indicator   Surface plate
Height gauge   V-blocks
Micrometer (external)   Vernier
Micrometer (internal)   Vernier protractor

Metal joining

Adhesives   Rivetting
Brazing   Soldering
Gas and electric welding   Spot welding
Pop-rivetting

Plastics

Adhesives   Injection moulding
Bending   Polishing
Carbon fibres   Sawing
Drilling   Turning
Glass resins   Vacuum forming
Hot-wire cutting   Welding

Safety

Electrical safety	Lubricants
Eyeshields	Projection of hands
Fumes	Rag
Guards	Safe clamping methods
Handling tools	Swarf and burrs

Tool and fitting

Centre punch	Scrapers
Clamps	Scribers
Files	Taps and dies
Hacksaw	Vices
Hole cutters	

Woodworking

Adhesives	Finishing
Bandsaw	Hand tools
Boring	Machining
Drilling	Turning

DATA SHEET 3

MANUFACTURING PROCESSES AND DESIGN ELEMENTS

These checklists are written for students of engineering design and engineering product design. Familiarisation with everything listed may take considerable time and the lists are intended as a reminder.

The headings should be interpreted broadly: thus, for example, gears – obviously there are many types; seek information on the whole range.

Extend the lists for special areas of design interest.

Design elements

Abrasives	Bellows
Actuators	Bolts, nuts and washers
Air compressors	Chain and belt drives
Anti-vibration devices	Clutches

Couplings
Dowel pins
Electric motors
Expanded metal materials
Gearboxes
Gears
Grub screws
Hank bushes
Hoses
Hydraulic circuit components
Inserts
Intermittent motion devices
Journal bearings
Locking devices
Magnets
Metal fasteners
Microswitches
Oil pumps
Optical devices (lenses, prisms, etc.)

Photo-electric cell
Pipe fittings
Pneumatic circuit components
Pulleys
Relays
Roll pins
Roller bearings
Seals
Self-tapping screws
Slotted angle
Split pins
Spring washers
Springs
Stepper motors
Thermostats
Timers
Water pumps
Wire ropes

Forming and treatment processes

Annealing
Case hardening
Crimping
Die casting
Drawing
Explosive forming
Extruding
Fettling
Forging
Hardening and tempering
Injection moulding
Investment casting
Knurling

Magnetic forming
Metal deposition
Nitriding
Normalising
Pressing
Rolling
Sand casting
Shell moulding
Sintering
Stretching
Swaging
Tumbling
Weathering

Machine processes

Bending
Boring
Broaching

Cutting (guillotine)
Cylindrical grinding
Die sinking

Drilling	Planing
Electrochemical machining	Punching
Engraving	Reaming
Flame cutting	Routing
Hobbing	Sawing
Honing	Shaping
Jig boring	Slitting
Jigs and fixtures	Slotting
Lapping	Spark erosion
Laser machining	Spinning
Milling	Surface grinding
Multi-purpose machine tools	Threading
Nibbling	Turning
Piercing	Ultrasonic machining

Metal finishing

Anodising
Chromium plating
Copper and Tin plating
Enamelling
Galvanising
Nickel plating
Paint spraying
Plastic coating
Precious metals
Tin plating

Quality control

Automatic inspection techniques
British standards
Checking procedures
Comparators
Limit gauges
Material testing
Non-destructive testing
Pneumatic gauging
Shadowgraph techniques
Surface finish measurement

DATA SHEET 4

COST ESTIMATION FORM

This is a blank form of the type described in Section 1.5 (costing). With a little practice it is simple to use and will help promote an awareness of costs, even when used for simple projects.

To use this form make sensible estimates for labour, overheads and profit margin and use estimated times and material costs from your own project.

Cost estimation is involved with every commercial design project; practice with simple estimating exercises helps to develop a useful instinct for the relationship between design proposals and cost.

| COST ESTIMATE | PROJECT | | |
| DATE | | | |

LABOUR	HOURS	CHARGE 36 HR/WEEK	COST
SKILLED			
SEMI-SKILLED			
UNSKILLED			
TOTAL LABOUR COST			

OVERHEADS   COST AT [   ]%   SUB-TOTAL

PROFIT   COST AT [   ]%   SUB-TOTAL

PURCHASES

MATERIALS
BOUGHT-IN ITEMS
SUBCONTRACTING

TOTAL PURCHASES
ADMINISTRATION CHARGE AT 10%
TOTAL CHARGE TO PROJECT

SUB-TOTAL

CONTINGENCY   COST AT [   ]%

COMMENT   TOTAL

DATA SHEET 5

LECTURING GUIDELINES

Designers are constantly involved in presenting design proposals and technical data etc., beginning during training and continuing as professional designers at seminars, exhibitions and conferences. Developing good presentation techniques has obvious advantages and this data sheet will help with early practice.

Technique

- Use good English but do not change your natural speaking voice.
- Time your lecture properly; do not speak quickly to get it all in.

- Do not speak to the board or screen; always address your audience.
- Do not 'mumble' when deciding what to say next.
- Avoid annoying repetitions; words such as er!, you know!, or O.K.! at the end of each sentence.
- Avoid slang words unless used for impact or effect.
- Avoid ambiguous statements such as '... and anything like that!'.
- Make sure that you can be heard at the back of the room.
- If your subject area is vast (for example, 'bearings') introduce which aspect is to be the focus of your talk at the beginning of the lecture.
- Try to keep note reading to a minimum by using reminder cards, widely spaced notes in large print, or key words.
- Avoid stating the obvious such as '... as you can see this is a square component...'.
- Do not make uncertain statements which cast doubt on your knowledge of a subject. When in doubt – leave it out!.
- Occasional jokes or amusing comments are useful but don't overdo it'
- Do not use silly remarks to overcome nervousness at the beginning of a lecture. The feeling usually disappears as you begin to speak.
- Avoid overrunning your lecture time; audiences lose concentration and start to fidget (even if they are still interested in what you have to say).
- Try to avoid lecturing while handout material or samples are being distributed.

Visual aids

- Prepare your sketches on paper before using the board or overhead projector (OHP) wherever possible.
- Make sure that screen size is not exceeded by slides or OHP projection. Check everything before your lecture.
- If you refer to small detail make sure that it can be seen.
- Use a pointer – not a finger!
- If a slide extension lead is not available, arrange for another person to operate a slide projector.
- Have all slides and OHP material ready and in the right order.
- Do not hold up small mechanisms or models and expect everybody to be able to see them.
- Make sure that slides and photographs are clear and easily understood.

Displaying Technical Data

- When lecturing about a process (e.g. extrusion of metals) show the process before describing it. Technical information in a lecture is incomprehensible if the process is not demonstrated initially.

- Whatever the presentation, make sure that wording on graphs and charts is large enough to be seen at the back of the lecture room. Too often we see data presented with lettering on x- and y-axes, or along curves on charts, in miniscule printing; the confusion is not avoided if unreadable data is explained – the information is largely forgotten immediately an audience listens to the next sentence.
- Explain carefully any unfamiliar technical words.
- Describing mechanisms without models or diagrams doesn't work!
- If tables of numbers are to be used make sure that the data you wish to discuss is easily read and understood. Use colours to pick out information if necessary.
- Do not remove diagrams too quickly: your audience will need a little time to absorb information.
- Be prepared to answer questions outside the basic lecture material by reading your subject thoroughly.
- If you suspect that part of your lecture is boring, find ways to make it more interesting by innovative use of the visual aids at your disposal.
- Do not 'gloss-over' vital comments relating to an apparatus or process. Allow important statements to sink in.

DATA SHEET 6

DESIGN EXHIBITION – ORGANISATION CHECK LIST

The term 'design exhibition' is used here to cover all types of display in which a designer may be involved, beginning with project presentation in the early years of training, organising exhibitions for degree shows and mounting or supervising exhibitions in the commercial world. Making sure that everything is organised can be a formidable task for people new to this activity. This check list covers most items which need attention and the reader is invited to extract from it all that may be useful to a particular type of display.

General

- Establish organising committee (six months early).
- Appoint a member of staff to organising committee.
- Book exhibition venue.
- Check insurance cover for exhibition and exhibits.
- Produce exhibition layout scheme for electricity supply planning.
- Plan exhibition 'centre' area – platform, flowers, college literature, symbolic exhibit and information desk.
- Order display boards, tables and chairs.

- Plan visitors' rest area and cloakroom.
- Arrange for storage of equipment and exhibits when exhibition is finished.
- Organise post-exhibition information centre.

Advertising

- Design posters and tickets (two months prior to exhibition).
- Write an information sheet.
- Send exhibition literature to:
  —local and national press
  —professional institutions
  —colleges and schools
  —local industry and potential employers
  —other departments within the college
  —selected visitors.

Safety and Security Arrangements

- Send exhibition details to safety officer.
- Check statutory rules concerning:
  —spacing between stands
  —exit notices and access to exits
  —fire equipment and alarms
  —ventilation
  —maximum number of visitors (relative to exhibition floor area)
  —first aid service and equipment.
- Order direction and exit notices.
- Provide exhibitors with instructions for guarding machinery and equipment.

Technical Support

- Workshop support – electronic and mechanical repairs.
- Maintenance support – lighting and services.
- Visual aid technician – films, video, slide projectors, background music etc.
- Exhibition photographer.

Services

- Exhibition details to domestic services department.
- Porters for door duties.
- Access and direction notices for toilets.
- Refreshment service.

- Litter bins.
- Access to telephones for visitors.
- Duty officers for information centre.

Individual Exhibition Stands

- Table for drawings and reports.
- Table for projects and equipment.
- Table covers.
- Display boards.
- Demonstration area for mobile projects.
- Chair.
- Decorations (plants etc.).
- Note pad.
- Special equipment e.g. dust or swarf trays, floor covers, splash guards etc.
- Course news sheets.
- Brief information sheet giving personal details of student exhibitor.
- Contact the Patent Office in good time to get temporary cover for patentable items to be exhibited.

DATA SHEET 7

PROJECT SPECIFICATIONS

A specification is a set of agreed general intentions which define a design proposal; it contains details of the aims and objectives of the product, design requirements, estimates of costs, completion dates and all other data that can be specified at the beginning of a project; it may even contain a clause agreeing to the provision of a revised specification at some future date. In the commercial world the contents of a specification may vary from a simple letter of intent to a document the size of a book. Whatever form it takes the data listed here will apply.

Rules for Writing a Specification

1. Make clear, concise and unambiguous statements.
2. Keep the number of words to a minimum required for full understanding by the customer, i.e. keep to the point!
3. Try to get an expert in your field of interest to read the specification before presenting it to a customer.
4. Wherever possible ensure that a feasibility study, estimate of material costs and a proposed working schedule are completed before writing a specification.

5. When specifying time estimates, make sure that sufficient margins are allowed for activities outside your personal knowledge.
6. Avoid commitment to a fixed cost estimate and positive completion date, unless these can be achieved without reasonable doubt.
7. Get written or signed approval for any modifications to an agreed specification.
8. Specified time estimates should allow for model constructions, trials, completing drawings and reports – activities which are often underestimated.
9. Include proposals for safety aspects where appropriate.

Working to a Specification

1. Make sure that every word and instruction is thoroughly understood.
2. Never modify a received specification without a written or signed agreement with your customer.
3. Frequent reference to the specification should be made throughout the duration of a project, to ensure that design decisions and concepts relate accurately to the stated objectives and instructions.

DATA SHEETS 8–22

PROJECT SPECIFICATIONS

These specifications extend those given in Section 1.7. The primary objective of including them is to suggest ideas and sources of project subjects.

Readers who may wish to use these specifications as a basis for a project are advised that the wording needs to be modified with due regard to the completion date, availability of finance and studio facilities, and with particular regard to the *quality* of the finished product – which can vary from a working or demonstration model to a near perfect prototype.

Sources of projects, other than those devised by the author, are identified with each specification.

PROJECT SPECIFICATION – MOUTH-CONTROLLED ROBOT ARM

Source: Hospital for handicapped children.

Project: Mouth-controlled 'robot' arm, intended for teenage children suffering from some form of paralysis. The device will provide a means whereby patients can play games, such as chess.

High degrees of positional accuracy may be avoided by designing the chessmen and board (or pieces used in other games) in conjunction with the shifting mechanism.

Patients are capable of sucking or blowing a number of mouthpieces (arranged in a similar manner to pan-pipes) to direct the arm movements.

PROJECT SPECIFICATION – HALF-STEP

Title: Half-step for disabled people

Source: Rheumatic Arthritis Association

Description: A requirement exists for a walking stick/half-step for people suffering from acute arthritis.

These patients are unable to raise a foot high enough to negotiate pavement steps and the short steps often placed at the entrance of shops. Many, however, are capable of raising a foot to about half the height of a step and would benefit from a device to assist them.

Details:

Weight: 2.25 kg maximum

Walking stick height: adjustable to individuals

Stability: must feel safe – users are mainly elderly people.

Mechanism: half-step lowered and raised easily (often using hands affected by arthritis).

Features: device must be provided with non-slip type crutch tips and a suitable hand grip;

a safety device will be needed to prevent accidental release of the half-step;

step to be located away from the leg in the walking mode;

a smooth, attractive product is essential;

to be produced in batches of 500.

Note: it is suggested that composite aluminium/foam materials are investigated for the platform. These are very light and probably rigid enough if provided with extra strengthening at the edges.

DATA SHEET 10

PROJECT SPECIFICATION – ILLUMINATED VISUAL AID

Source: Lecturer – nursing degree course
Project: An illuminated teaching aid which provides a step-by-step graphical presentation of the path taken by neurotransmitters – from dendrite receptors via brain cell body, axon and synapses to nerve network destination.

  Note: Obviously, contact with the originator of this project must be made before this project can begin. It is included here to advise that projects can be devised from visual aids required by many activities unassociated with engineering.

DATA SHEET 11

PROJECT SPECIFICATION – VISUAL AID FOR BEAM CURVATURE

Source: Lecturer – engineering science
Project: Design a visual aid which demonstrates beam curvature under various load and support conditions.
Movement of the load or supports must result in exaggerated deflections of the beam to those calculated mathematically.
The placings of supports and loads need not be infinitely variable but the beam deflections (curvature) must be clearly seen by an audience.

DATA SHEET 12

PROJECT SPECIFICATION – MECHANISM VISUAL AID

Project: Visualising the geometric relationship between the input and output motions of a mechanism is often difficult – and sometimes impossible – particularly if the mechanism links more than two components that do not function as a pure gear ratio or which operate in more than one plane. The best method of learning about such mechanisms is to see and handle them in the form of accurately engineered models; this will establish a lasting 'mental image' and provide experience in the mechanism function. The project aims to provide hand-operated visual aids to assist with the teaching of mechanism design.
Mechanisms may be copied from existing machines, apparatus or illustrations.

The model must be of robust construction and large enough to be demonstrated from the front of a class of students (who will also handle the mechanism and activate it by a suitable input control).

Two useful innovations are (depending on the type of mechanism):

1. a means of adjusting one component to demonstrate changes in the output geometry;
2. a means of plotting the output geometry on paper (or any other means of display).

DATA SHEET 13

PROJECT SPECIFICATION – STIFFNESS COMPARATOR

Project: Design an instrument that can be used to demonstrate stiffness in bending for several specimens of different materials.

Each specimen (manufactured to the same length and cross-sectional area) is to be clamped at one end and subjected to an identical force (weight) at a fixed distance from the free end.

Loading each specimen results in a certain amount of deflection at the free end and measuring this deflection provides a comparison of stiffness. Deflections of the specimens are to be magnified in a manner which can be displayed on a suitable scale (or other means) and easily read. It is advisable that a carrier weight is placed on each specimen and a calibration adjustment applied to the readout (for zero setting) before applying the deflection force.

A useful innovation would also compare the stiffness of specimens which have the same dimensions as solid specimens but which have various cross-sectional geometries.

DATA SHEET 14

PROJECT SPECIFICATION – BISCUIT TESTING DEVICE

Source: A food research department.
Project: Design an instrument for measuring and recording the breaking strength of biscuits.

Biscuits of one particular size can be manufactured using slightly different ingredients and/or cooking specifications. This produces variations in the physical properties of the finished product. The instrument proposed will accept a biscuit into a suitable nest. It will rest on a flat 'ring' of approximately 8 mm width (not a critical measurement, since it will be a constant feature for each biscuit tested). Testing the biscuit will be achieved by loading at the centre with a steel ball of approximately 10 mm diameter. An increasing load will be applied to the biscuit until it breaks. The force at breaking point will be recorded with the biscuit code number.

DATA SHEET 15

PROJECT SPECIFICATION – BENDING AND FORMING TOOL

Source: Lecturer – engineering design.

Description: A requirement exists for a bench- or vice-mounted device which can be used for bending metal sheet and strip or for cropping and bending metal wire.

Data: Maximum width of sheet metal: 5 cm
                Maximum gauge of sheet metal: 20 SWG
                Maximum diameter of wire to be cropped or bent: 12 SWG
                The tool must be hand-operated
                Ancillary fittings such as back-stops and bending pillars, or other aids to using the tool should be included in the design.

DATA SHEET 16

PROJECT SPECIFICATION – UNIVERSAL CLAMP

Description: Clamping of components is necessary for a variety of activities in craft work. Examples are gluing, soldering, working of thin metals, wire work, tacking and pin-nailing, painting, hand-drilling etc. A clamp is required which will provide for many of these purposes either for the model maker or for art workers or jewellers.

Data: The device should be versatile, relatively simple in concept and exhibit a degree of elegance in its design.
                If the product is to be aesthetically pleasing, care must be exercised in the degree of versatility put into the design.
                Attachments would be acceptable and these should be stored in a compartment of the product.
                A product aimed at one particular craft (such as model aircraft work) would also be acceptable.

DATA SHEET 17

PRODUCT SPECIFICATION – SPECIAL-PURPOSE LATHE

Description: This machine is intended for use in the design studio workshop and will ease the workload on existing lathes. Its function will be confined to producing small and simple components for general use in project work – typically stand-offs, spacers, 'hank' bushes, small washers, small components for models.

Design data: Maximum diameter of components 10 mm
Maximum longitudinal movement of toolholder 50 mm
Motor driven with choice of two speeds
Operation by hand (no automatic tool feed)
Designed with suitable stops for repeated operation
Simple and quick setting of machine and cutting tools is of paramount importance.

DATA SHEET 18

PROJECT SPECIFICATION – HIGH-SPEED DRILLING MACHINE

Description: A special drilling machine is required for the design studio workshop, to be used for drilling small holes in sheet materials (maximum thickness 1.5 mm).

When a small hole-drilling operation occurs at present, students are obliged to resort to an unsatisfactory method of holding the drill in a pin-chuck which, in turn, is held in a normal drilling chuck. This results in the drill running out of true and damage to the pin-chuck as attempts are made to grip the drill with enough force to retain it. Additional problems are incorrect drilling speed and complete lack of 'feel' for the drilling operation, resulting in broken drills.

Design data: Drill sizes – 1 mm diameter maximum
Bench mounted
Lightweight and sensitive drilling action
The possibility of a safety device to prevent overload on the drills should be examined
The general structure should be solid and any vibrations from the driving motor isolated from the drilling operation.

DATA SHEET 19

PROJECT SPECIFICATION – ENGRAVING MACHINE (LINEAR SCALES)

Description: A small engraving machine is required which is designed specifically for engraving linear scales of various lengths and line spacings. It is intended that scales engraved on this machine are produced far more quickly than those produced (on a one-off basis) on the conventional engraving machine.

It must be possible to vary the length of scale and to carry out the engraving process without using a range of master scales.

Design data: Scale lengths: 50–250 mm

Scale widths: 10–20 mm

Length of lines: 2–6 mm

Facilities for including scale numbers should be investigated during the design study. If this feature can be achieved using simple attachments or adjustments, it should be incorporated.

Engraving operation to be controlled by hand.

DATA SHEET 20

PROJECT SPECIFICATION – ENGRAVING MACHINE (CIRCULAR SCALES)

Description: A bench-mounted engraving machine is required which is designed specifically for engraving circular scales of various diameters or to engrave circular arc scales on discs. It is intended that this machine is quick and easy to use, particularly when the need is for one scale only. It must be possible to vary the arc length of the scale up to full diameter.

Design data: Engraving surfaces: edge and bevel.

Disc diameters: 40–80 mm

Disc thicknesses: 1.5–15 mm

Facilities for including scale numbers should be investigated during the design study. If this feature can be achieved using simple attachments or adjustments, it should be incorporated.

Engraving operation to be controlled by hand.

DATA SHEET 21

PROJECT SPECIFICATION – PADDLE BOAT

Description: Paddle boats are a common sight on the lakes and ponds of amusement parks. Their general form varies from place to place and usually has the image of a boat, car or 'raft'; but for many years there has been little effort to produce anything new.

You have been retained by a company – whose normal business is fairground equipment – to design a paddle boat which has an entirely new, exciting and futuristic image.

DATA SHEET 22

PROJECT SPECIFICATION – CHILDREN'S ROUNDABOUT

Description: An effort is being made by several district councils to improve safety standards of equipment in childrens' playgrounds and also to introduce new types of apparatus. One proposal, which has been favourably received, is for an apparatus which is driven by a push–pull action by the child, resulting in rotation of the apparatus and the child, about a central axis.

The device must be attractive in appearance and suitable for children up to eight years of age.

The quantity required each year is 500.

DATA SHEET 23

A RECOMMENDED PROCEDURE FOR PROJECT PLANNING

It is suggested that this procedure is first practised on a relatively uncomplicated project, such as a child's toy, garden tool or something similar in degree. Even with a small product the advantages of using this project-planning approach are self-evident as the study progresses.

1. Specification Analysis

Project Review

A careful examination of the project objectives.

Survey of Design Parameters

Careful reading of the specification – underlining and noting everything that may influence the design (including any potentially useful ideas that emerge).

Specification Study

For each main element the question is posed 'is there anything we do not know concerning …'? Answers are now sought to any queries generated by this question – in addition to any other queries not yet settled.

2. Review of Priorities

A formal discussion on priorities.

3. Study of Basic Concepts

Includes ideas emerging from:
- unconventional solutions
- fundamental principles
- new materials, processes and techniques
- visualisation of product in operation
- human and environmental inputs to the project
- matters detrimental to the life or efficiency of the product.

4. Summary Listing of Design Study Conclusions

DATA SHEET 24

METHOD OF WORKING AND REPORTING

In practice designers develop their own preferred methods of working and reporting; these activities may be influenced by the nature of the product, working environment and company policy etc. Whatever approach is ultimately adopted in managing a project, the procedures outlined in this data sheet will have to be addressed at some stage.

The following list of design activities outlines an approximate sequence of events involved in a full project management programme:

- project planning

- design study
- first estimation of costs
- production of drawings
- revised estimate of costs
- manufacture of a prototype
- prototype trials and possible modifications
- firm estimate of costs
- writing a formal report.

In practice there is an overlapping of these activities; ideas developing at each phase often modifying earlier thoughts about design details, costs, prototype manufacture and the organisation of the project. An understanding of the subtleties with which design activities integrate is an *evolutionary* process which develops rapidly with experience and which begins with the first complete projects undertaken on a design course.

The second phase in the above list is identified as *design study*. Opinions vary on what constitutes a design study; in the context of the above programme, it is defined as the period immediately following the *choice of scheme* emerging from a project-planning study. Obviously there are many instances where a project *begins* with a firm concept (the choice of scheme already exists); examples are new inventions, modifications to existing products, concepts demanded by a rigid specification, etc. With projects of this nature project planning is synonymous with design study and the same thorough approach is essential.

Students undertaking their first project should not be disappointed if estimates of costs and time schedules prove to be inaccurate. Estimates are difficult initially, but become easier with experience; like every other skill, it needs practice. A designer working in industry is constantly making rough estimates for proposed projects – and it is surprising how familiarity with particular products enables the designer to quote for similar projects with a relatively small contingency margin.

Other matters of interest are:

1. Once a final design concept has been chosen, there are advantages in making a large layout drawing of it (not to be confused with a general assembly drawing). Modifications to components and assemblies usually accompany efforts to engineer the basic concept of a design and a large drawing helps with visualising the final product. 'Thumbnail' sketches should be avoided; these make communication difficult and tend to disguise or oversimplify mechanical complexity. The ability to produce neat line sketches is a basic skill which should be practised at every opportunity.
2. It is advisable that final details are worked out on a formal drawing, even if a complete set of formal drawings is not required. It is usually necessary to

produce several views and a number of formally drawn details to establish the basic product design.

3. If a design report is specified it normally consists of (and in this order):

(a) Cover – with name, course and date on the outside.
(b) Title page.
(c) Index of contents (with page numbers).
(d) Specification (and any relevant data).
(e) Summary – this discusses briefly the final design and costs. It describes how the specification has been met and often includes a pictorial presentation or photograph of the final product. A summary is meant to be read by all concerned with the project – including people who are not designers – and this should be reflected in the wording. The summary will also include conclusions and recommendations.
(f) Report – this comments on matters relating to the final design concept. It includes all information useful to the customer and those who will be concerned with manufacture and field trials. Typical contents are:

    (i) A full set of drawings, which must include
- all instruction necessary for component manufacture;
- material specifications for every component;
- tolerances on all dimensions, with due consideration to plating or other finishing processes;
- complete instructions for assembly, including special notes on quality checking where necessary;
- approved drawing numbering system, with due regard to drawing modification procedure.

    (ii) Comments on main design features such as safety, ergonomics, maintenance, etc.

    (iii) Results of experiments or proving tests and data important to the basic design concept (where appropriate or instructed by the specification).

    Note: Comments on rejected design concepts should be excluded except where essential to the understanding of some aspect of the report.

(g) Operating instructions.
(h) Appendices: For items such as manufacturers' catalogues, correspondence and other data relevant to the report but only required for reference purposes.

MODEL TECHNIQUES – SUBJECTS FOR FURTHER READING

The subject 'model techniques' in design literature is extensive and extremely broad. There is considerable educational value in occasional visits to design libraries for the purpose of surveying the subjects listed in this data sheet:

scale effects using models
the use of models in research
models and communication
ergonomic applications of models
models in the drawing office
aesthetic design models
economic advantages of using models

model as aids to structural design
models and mechanism design
modelling on a computer
using models – advantages and
   limitations
case studies in model techniques.

Glossary

Note: Word definitions are confined to engineering terminology.

academic	scholarly
accelerate	increase speed
accumulate	increase or amass
acquisition	gaining
activate	start or turn on
actuator	a device which causes movement of something
acute	sharp — less than a right angle
adapting	making it suit a purpose
adequate	sufficient
adhesive	glue-like substance
adjacent	near or close to
adjustable	able to be moved or reset
aesthetic	artistic, appealing, having beauty
algae	a plant – of which one type grows in static water
aligned	lined up with something (often making parallel)
alleviate	lessen, ease
amplify	intensify, increase, expand
analyse	examine, study in detail, investigate
ancillary	additional in some way
annual	yearly
appraisal	valuation, assessment of worth
appropriate	specially relevant or suitable
articulated	movable joint (such as an elbow or ball and socket)
assemble	join together
assertion	a confident statement
assessment	examination of value or forming an opinion
assimilate	absorb into the mind
attributes	talents, abilities, skills
audible	able to be heard
automatic	controlled by machine
backlash	slight looseness between components

bearing	load-carrying component which is subject to friction
bizarre	impossible from an engineering or economic viewpoint
brazing	joining together using a hard soldering process
buckling	the bending of components and sheets under the influence of forces
budget	allowable expenditure
buffer	spring-loaded device used to absorb the shock caused by a blow or force
calibrate	set up to a given mark or value
capacity	amount something can hold, maximum volume or energy output
catalyst	stimulator, influencing something to happen
chamfer	a flat 45° bevel along an edge
clutch	a mechanical device for connecting two components while one, or both, are rotating
commercial	in business or industry
commission	an order to action some kind of work
compact	closely fitted, small, without wasted space
comparative	compared with something else
compatible	in harmony with
component	part or piece
compromises	agreed variations in previous ideas
concept	inventive idea
concession	consent to do something not previously agreed (such as new ideas, modifications, costs)
concise	brief, to the point
conformed	complied with, obeyed, fitted in with
conservation	care, protection, preservation
constitute	make up, include, contain
construction	built-up assembly of parts
consultation	joint discussions
contaminated	polluted, dirty, infected
contingency	allowance for unforeseen events
converted	change from one state to another
coordination	proper control
corrosion	rusting, wearing away through electrical, gaseous or chemical activity
criteria	rules, standards or data by which something is assessed
crucible	vessel in which metal ores are melted
culminating	ending or finishing
cycle	events happening in a repeating order
data	facts, relevant information
define	describe accurately
degree	proportion, measured quantity, extent
detectable	able to be found out
detrimental	harmful in some way, undesirable
deviation	changing from a set mark, value or course
device	invention, gadget, mechanism
devise	think up, invent
diagnostic	inquiring by concentrated study
diligent	careful or observant
dimension	size

diminish	lessen or shorten
discretion	judgement, individual choice
dismantle	take apart
dispatch	send off
dispense	eject in controlled quantities
displacement	amount of movement
durable	long-lasting
dynamic	energetic
eccentric	off centre
economic	not wasteful
edible	able to be eaten
efficacy	how well an effect is produced
efficient	highly capable
elapsed	passed by
emerge	come out
enhance	intensify, make more attractive
environment	surroundings, habitat
ergonomic	designed to fit the individual
estimate	approximate cost
evaluation	judgement
evolves	gradually develops
exclusive	special only to one thing
extensive	large, widespread
extrapolate	conclude something from facts already known
extruded	forced out through a small opening (like toothpaste)
fabricate	manufacture, put together
facet	one particular view (e.g. one face of a many-sided crystal)
facilitate	help by making something easier
fatigue	tiredness
feasibility	possibility to achieve
feeler gauge	a tool for measuring thin gaps between components: graded spacing pieces range from 0.025 mm to 0.625 mm
fluctuations	repeated changes
focus	pay special attention to a particular thing
formal	accurate, proper, not free-hand
formidable	hard to do, very demanding
frequency	the number of repeats of something
friction	resistant force felt when objects rub together
function	purpose or activity
hopper	container of components or materials (where the contents come out under some form of control)
hybrid	emerging from more than one source
hydraulic	using a fluid to transmit a force
illustrated	made clear, explained with pictures or drawings
impaired	weakened, unsatisfactory, having smooth control spoiled by something
imperative	having high priority, urgent, unquestionable action
implement	start, put into action
implication	outcome, effect of

impurities	substances which contaminate
inclusion	small particle of other material
incorporate	include within something
increment	a measured increase in distance, angle or rotation
indexing	moving in steps (increments) under some form of control
inertia	reluctance to move
inevitable	certain to be or happen
infinitely	without boundary, unmeasurable
infrastructure	existing framework or whole structure
infringe	get in the way of
ingenious	imaginative, highly skilled, brilliantly inventive
inhibitor	substance which prevents the growth of algae in water tanks
initial	the first
innovation	invention of new concepts
insert	component that has been put into another
insight	instinctive understanding
integrate	combine or blend
intermittent	occurring at intervals of time
interpret	explain, make clear
intractable	resisting a solution, stubborn
intuition	instinct, inner sense
invert	turn upside down
involute	the curve or geometric (ideal) shape of spur gear teeth
kinematics	study of motion without consideration of the force causing the motion
location	place or position
loci	points through which a curve is plotted
logical	correctly reasoned
maintenance	servicing, making repairs, looking after
manipulation	influencing things for a special purpose
manual	operated by hand
mechanism	an arrangement of components which are designed for a special function
metering	measuring something – usually by reading a dial
minor	of very small consequence
modify	adjust, change, vary, re-work
motivate	provide the will to undertake something
neural	about nerves or the nervous system
nominal	not substantial
objective	target, main intention
optimum	the best possible
ore	mineral of a metal before smelting
orientate	control the direction, move to locate
outrigger	end support of a long component or assembly
overheads	costs due to non-productive items such as management, administration, buildings, maintenance, etc.
parameters	all the things which relate to what is being considered, the features of something
patent	legal protection of an invention
pawl	something which catches on to a component

peripheral	around the edges of, surrounding
perspective	view, 'looked at from this angle'
pertinent	relevant, suitable
phase	stage of development, juncture reached
philosophy	wisdom and knowledge of the causes of things
physiotherapy	treatment by physical methods such as exercise, massage, heat, electricity
pivot	component around which another component rotates
pneumatic	using air to transmit a force
porous	usually – full of tiny holes
potential	capable of existence – mentally or physically
practicable	satisfactorily feasible
precedent	same or similar happening in the past
precision	accuracy
preconceived	already thought of
premium	best value
prestige	esteem enjoyed by a person or product
primary	first in importance or first to happen
priority	first choice, top place
profile	outline
progressive	gradually moving to achieve something
prototype	the first or early versions of a product
quality control	inspection and examination of production procedures, manufactured items, materials, performance etc.
quantum leap	sudden jump from one state to another
questionnaire	list of questions – usually for collecting statistical evidence
ratio	proportion, in relation to
reamed	finished to an exact size by a special tool
receptive	quick response to an idea
recipient	one who receives something
reciprocate	move back and forth
reinforced	strengthened
relatively	as compared with the subject under consideration
relevant	relating to do with the subject considered
reliable	dependable, of good quality
replica	a copy
reservoir	a place where things are stored
resonance	rebounding or echoing vibration or sound
resource	provider of things – materials, ideas, skills, know-how etc.
review	re-examine, look at again
roll pin	small cylindrical spring tube which provides a dowel function using drilled holes only (without the need for reaming)
safety interlock	a mechanism or device which operates when a product malfunctions – preventing harm or danger
schedule	list or programme of target dates for completing phases of a project
schematic	outline of an idea or concept
scuffing	damaging the surface, scoring and scratching
seasoning	drying, weathering
sectioned	drawn as if cut through at convenient places to show what is inside

seminar	a discussion group
sequence	set of events following each other
serendipity	accidental discoveries by a person skilled enough to recognise their value
shim	very thin metal sheet or strip
simulate	act like, be similar to
smelted	metal obtained from its ore by heating
spacer	a component that is used to keep other components separated by a fixed distance
specification	a description stating the customer's requirements for a product (usually a document)
specify	describe in detail
specimen	a sample, a part (from which the whole can be assessed)
spectrum	features or factors spread out and easily seen or visualised
spontaneous	impulsive, without sufficient thought
static	not moving
subassembly	a number of associated components collectively forming one part of the main assembly (the product)
subcontract	order from another company
subsequently	following on, afterwards
substitute	alternative, used in place of something else
suffice	to be adequate, to be enough
survey	examine, look at, get acquainted with
swage	bend out by striking with a suitable tool
technique	method of achievement
tenacious	clinging, persistent
theme	topic
therapy	medical treatment
thermal	concerned with heat
thermocouple	a device which measures temperature differences
tolerance	the amount by which a dimension is allowed to vary
torsion	twisting
traditional	customary
transducer	a device which receives some form of power and transmits it in another form (e.g. sound waves into electrical signals)
transmission	passing on or transferring
transverse	lying across (usually at right angles to)
ultrasonic	concerned with sound waves
uninhibited	free thought, without being influenced by conventional thinking
unique	being the only one, like nothing else
utilise	make use of
variable	able to be changed – in speed, voltage or any other controllable activity
venture	an undertaking
versatile	able to be used in many ways
viable	having a good chance of success
virtually	almost, as near as possible
visualise	make a mental image

Index